STUDENT SOLUTIONS MANUAL

TO ACCOMPANY

A MODERN INTRODUCTION TO DIFFERENTIAL EQUATIONS

Henry Ricardo
Medgar Evers College of the City University of New York

HOUGHTON MIFFLIN COMPANY BOSTON NEW YORK

Editor-in-Chief: Jack Shira
Sponsoring Editor: Lauren Schultz
Editorial Associate: Marika Hoe
Senior Manufacturing Coordinator: Jane Spelman
Marketing Manager: Ben Rivera

Printed in the U.S.A.

ISBN: 0-618-04241-5

123456789 – MA – 06 05 04 03 02

Contents

Chapter 1

Introduction to Differential Equations

1.1 Basic Terminology

1. (a) The independent variable is x and the dependent variable is y; (b) first-order; (c) linear

3. (a) The independent variable is not indicated, but the dependent variable is x; (b) second-order; (c) nonlinear because of the term exp(-x)—the equation cannot be written in the form (1.1.1), where y is replaced by x and x is replaced by the independent variable.

5. (a) The independent variable is x and the dependent variable is y; (b) first-order; (c) nonlinear because you get the terms $x^2(y')^2$ and $xy'y$ when you remove the parentheses.

7. (a) The independent variable is x and the dependent variable is y; (b) fourth-order; (c) linear

9. (a) The independent variable is x and the dependent variable is y; (b) first-order; (c) nonlinear because of the term $e^{y'}$

11. The terms $(a^2-a)x\dfrac{dx}{dt}$ and $te^{(a-1)x}$ make the equation nonlinear. If $a^2-a=0$— that is, if $a=0$ or $a=1$—then the first troublesome term disappears. However, only the value $a=1$ makes the second nonlinear term vanish as well. Thus $a=1$ is the answer.

1.2 Solutions of Differential Equations

1. $y=\sin x$, $y'=\cos x$, $y''=-\sin x$; thus $y''+y=-\sin x+\sin x=0$.

3. $y = x^2,\ dy/dx = 2x$; thus $(1/4)(dy/dx)^2 - x(dy/dx) + y = (1/4)(2x)^2 - x(2x) + x^2$
$= (1/4)(4x^2) - 2x^2 + x^2 = x^2 - 2x^2 + x^2 = 0.$

5. $y = at^3 + bt^2 + ct + d,\ dy/dt = 3at^2 + 2bt + c,\ d^2y/dt^2 = 6at + 2b,\ d^3y/dt^3 = 6a,$
$d^4y/dt^4 = 0.$

7. $y = \ln x^2,\ y' = (1/x^2)(2x) = 2/x$; thus $xy' - 2 = x(2/x) - 2 = 2 - 2 = 0.$

9. $y = x^2/2 + (x/2)\sqrt{x^2+1} + \ln\sqrt{x+\sqrt{x^2+1}},$
$y' = x + (x/2)(1/2)(2x/\sqrt{x^2+1}) + (1/2)\sqrt{x^2+1} + (1/2)\left(1 + x/\sqrt{x^2+1}\right)/\left(x+\sqrt{x^2+1}\right)$
$= x + \sqrt{x^2+1}$; thus $xy' + \ln(y') = \left(x^2 + x\sqrt{x^2+1}\right) + \ln\left(x + \sqrt{x^2+1}\right)$
$= \left(x^2 + x\sqrt{x^2+1}\right) + \ln\left(x+\sqrt{x^2+1}\right) = \left(x^2 + x\sqrt{x^2+1}\right) + 2\ln\sqrt{x+\sqrt{x^2+1}}$
$= 2\left(x^2/2 + (x/2)\sqrt{x^2+1} + \ln\sqrt{x+\sqrt{x^2+1}}\right) = 2y.$

11. $y = \int_3^x e^{-t^2}\,dt,$ so that $y' = e^{-x^2}$ and $y'' = -2xe^{-x^2}$. Thus
$y'' + 2xy' = (-2xe^{-x^2}) + 2x\left(e^{-x^2}\right) = -2xe^{-x^2} + 2xe^{-x^2} = 0.$ (See Appendix A.4, statement (B), for the FTC.)

13. (a) The given equation is equivalent to $(y')^2 = -1$. Since there is no real-valued function y' whose square is negative, there can be no real-valued function y satisfying the equation.

(b) The only way that two absolute values can have a sum equal to zero is if each absolute value is itself zero. This says that y is identically equal to zero, so that the zero function is the only solution. The graph of this solution is the x-axis (if the independent variable is x).

15. If $y = \pm\sqrt{c^2 - x^2} = \pm\left(c^2 - x^2\right)^{1/2}$, then $dy/dx = \pm\frac{1}{2}\left(c^2 - x^2\right)^{-1/2}\cdot(-2x)$
$= \mp x\left(c^2 - x^2\right)^{-1/2}$ and $y\,dy/dx + x = \pm\left(c^2-x^2\right)^{1/2}\cdot \mp x\left(c^2-x^2\right)^{-1/2} + x$
$= -x + x = 0$. If $x > c$ or $x < -c$, then $c^2 - x^2 < 0$ and then the functions $y = \pm\sqrt{c^2 - x^2}$ do not exist as real-valued functions. If $x = \pm c$, then each function is the zero function, which is not a solution of the differential equation.

17. Since $y = at^3 + bt^2 + ct + d,\ y(0) = 1$ implies that $d = 1,\ y'(0) = 0$ implies that $c = 0,\ y''(0) = 1$ implies that $2b = 1$, or $b = 1/2$, and $y'''(0) = 6$ tells us that $6a = 6$, or $a = 1$. Thus the solution of the IVP is $y = t^3 + (1/2)t^2 + 1$.

19. We have $y = \left(e^{ax} + e^{-ax}\right)/2a$ and $y' = (ae^{ax} - ae^{-ax})/2a = (e^{ax} - e^{-ax})/2$, so that $y(0) = \left(e^{0} + e^{0}\right)/2a = 1/a = 2$ implies that $a = ½$ Noting that $y'(0) = \left(ae^{0} - ae^{0}\right)/2 = 0$ for *all* values of a, we conclude that $y(x) = e^{x/2} + e^{-x/2}$ is the solution of the initial value problem.

21. As was illustrated in Example 1.2.5, the velocity function is the derivative of the position function, so that we have $\dfrac{dx}{dt} = \dfrac{1}{t^2+1}$. Integrating both sides, we get $x(t) - x(0) = x(t) = \int_0^t \dfrac{1}{u^2+1}\,du = \arctan(t)$. (Also see equation (1.2.1).) For $t \geq 0$, arctan $(t) \leq \pi/2$, so that $x(t) \leq \pi/2$. (Look at the graph of the arctangent.)

23. The given information is that $v(0) = 0$, $v(30 \text{ seconds}) = v(30/3600 \text{ hours}) = 200$ mph, and $a(t) = C$, a constant. Now $a(t) = C \Rightarrow v(t) = \int a(t)dt = Ct + K$. Then $v(0) = 0 \Rightarrow K = 0 \Rightarrow v(t) = Ct$. Therefore $200 = v(30/3600) = v(1/120) = C(1/120) \Rightarrow C = 200(120)$ and $v(t) = 200(120)\,t$. Finally, distance $= s(1/120) =$

$$\int_0^{1/120} v(u)\,du = \int_0^{1/120} 200(120)u\,du = 12000(1/120)^2 = 5/6 \text{ mile}.$$

Alternatively, we can just draw the velocity curve (a straight line) and calculate the area under the curve, which is the area of a right triangle:

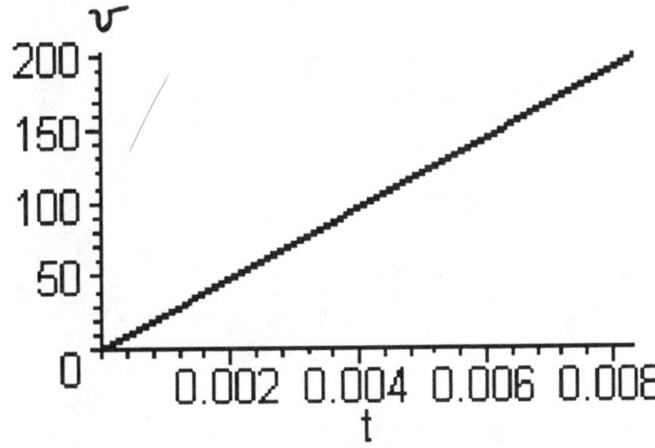

$$\text{Area} = \frac{1}{2}\left(\frac{1}{120}\text{ hours}\right)(200 \text{ miles / hour}) = 5/6 \text{ mile}.$$

25. Integrating each side of the given equation successively, we have $EI\,y^{(3)} = (-W/L)x + A$, $EI\,y^{(2)} = (-W/2L)x^2 + Ax + B$,

and $E\,I\,y'=(-W/6L)x^3+(A/2)x^2+Bx+C$. If we use the boundary condition $y'(0)=0$ in the last equation, we find that $C=0$. Integrating again, we get $E\,I\,y=(-W/24L)x^4+(A/6)x^3+(B/2)x^2+D$. Because $y(0)=0$, we get $D=0$. Finally, using the conditions $y(L)=0$ and $y'(L)=0$ in the equations for y and y', we get the algebraic equations $(-W/6)L^2+(A/2)L^2+BL=0$ and $(-W/24)L^3+(A/6)L^3+(B/2)L^2=0$, respectively. Solving these simultaneously for A and B, we find $A=W/2$ and $B=-WL/12$. Therefore the solution is $E\,I\,y=(-W/24L)x^4+(W/12)x^3-(WL/24)x^2=(-W/24L)\,x^2\,(x-L)^2$.

27. (a) If $y=\ln(|C_1x|)+C_2$, then $y'=C_1/C_1x=1/x$ for all values of C_1 and C_2 (with $C_1\neq 0$).
(b) Note that $y=\ln(|C_1x|)+C_2=\ln(|C_1|)+\ln(|x|)+C_2=\ln(|x|)+(\ln(|C_1|)+C_2)$ $=\ln(|x|)+C$, where $C=\ln(|C_1|)+C_2$.

29. Differentiating implicitly, we find that $xy'+y-y'/y=0$, so that $xyy'+y^2-y'=(x\,y-1)y'+y^2=0$, a first-order nonlinear equation, is a possible answer.

31. If $y(x)=c_1\sin x+c_2\cos x$, then
$dy/dx+y=(c_1\cos x-c_2\sin x)+(c_1\sin x+c_2\cos x)=(c_1-c_2)\sin x+(c_2+c_1)\cos x$.
If this last expression must equal sin x, then we must have $c_1-c_2=1$ and $c_2+c_1=0$. Adding these last equations, we find that $c_1=1/2$ and therefore $c_2=-1/2$. Therefore, the solution is $y(x)=(1/2)(\sin x-\cos x)$.

33. (a) We have $y=e^x=y'=y''$. Therefore
$x\,y''-(x+n)y'+n\,y=\;x\,y-(x+n)\,y+n\,y=0$.

(b) We have $y=\sum_{k=0}^{n}\frac{x^k}{k!}$, $y'=\sum_{k=0}^{n}\frac{k\cdot x^{k-1}}{k!}=\sum_{k=1}^{n}\frac{x^{k-1}}{(k-1)!}=y-\frac{x^n}{n!}$, and

$$y''=y'-\frac{n\,x^{n-1}}{n!}=y'-\frac{x^{n-1}}{(n-1)!}.$$ Therefore, $x\,y''-(x+n)\,y'+n\,y$

$$=x\left[y'-\frac{x^{n-1}}{(n-1)!}\right]-(x+n)y'+n\,y=x\,y'-\frac{x^n}{(n-1)!}-x\,y'-n\,y'+n\,y$$

$$=-\frac{x^n}{(n-1)!}-n\,y'+n\,y=-\frac{x^n}{(n-1)!}-n\left[y-\frac{x^n}{n!}\right]+n\,y$$

$$=-\frac{x^n}{(n-1)!}-n\,y+\frac{x^n}{(n-1)!}+n\,y=0.$$

35. (a) For any values of A and B, $x' = 3(A+Bt)e^{3t} + Be^{3t} = \{3(A+Bt)+B\}e^{3t}$
$= (3A+B+3Bt)e^{3t} = y$. Now $y' = 3(3A+B+3Bt)e^{3t} + 3Be^{3t}$
$= (9A+3B+9Bt+3B)e^{3t} = (9A+6B+9Bt)e^{3t}$ and
$-9x+6y = (-9A-9Bt)e^{3t} + (18A+6B+18Bt)e^{3t}$
$= (9A+6B+9Bt)e^{3t}$, so that $y' = -9x+6y$.

(b) The initial condition $x(0) = 1$ yields $1 = x(0) = (A+0)e^0 = A$, and $y(0) = 0$ gives us $0 = (3A+B)$, so that $B = -3$. The solution of this system IVP is therefore $\{x(t) = (1-3t)e^{3t},\ y(t) = -9te^{3t}\}$.

37. (a) Using the same reasoning found in Example 1.2.1, we see that $V_1(t) = V_1(0)e^{-ct} = V_0 e^{-ct}$.

(b) Let $T^*(t) = T^*(0)e^{-\delta t} + \frac{kT_0V_0}{\delta - c}\left(e^{-ct} - e^{-\delta t}\right)$. Then

$$\frac{dT^*}{dt} = -\delta T^*(0)e^{-\delta t} + \frac{kT_0V_0}{\delta - c}\left(-ce^{-ct} + \delta e^{-\delta t}\right)$$

$$= -\delta T^*(0)e^{-\delta t} - \frac{c}{\delta - c}\left(kV_0T_0e^{-ct}\right) + \frac{\delta}{\delta - c}\left(kV_0T_0e^{-\delta t}\right)$$

$$= -\delta T^*(0)e^{-\delta t} + kV_0T_0e^{-ct}\left(1 - \frac{\delta}{\delta - c}\right) + \frac{\delta}{\delta - c}\left(kV_0T_0\right)e^{-\delta t}$$

$$= -\delta T^*(0)e^{-\delta t} + kV_0T_0e^{-ct} - \frac{\delta kT_0V_0}{\delta - c}\left(e^{-ct} - e^{-\delta t}\right)$$

$$= k\left(V_0e^{-ct}\right)T_0 - \delta\left[T^*(0)e^{-\delta t} + \frac{kT_0V_0}{\delta - c}\left(e^{-ct} - e^{-\delta t}\right)\right]$$

= [using the result of (a)] $kV_1T_0 - \delta T^*$, so that T^* is a solution of the differential equation for T^*.

(c) $\lim_{t\to\infty} T^*(t) = \lim_{t\to\infty} T^*(0)e^{-\delta t} + \frac{kT_0V_0}{\delta - c}\left(e^{-ct} - e^{-\delta t}\right) = 0 + 0 = 0$. The number of infected cells decreases to zero.

Then we can write $w(t) = \frac{(1-k)}{k}Ae^{kat} + \frac{(k-1)A}{k} = \frac{(k-1)A}{k}\left(1 - e^{kat}\right)$.

39. Suppose $y = y_{GR} + y_P$. Then
$x^2y'' + xy' - 4y = x^2\left(y''_{GR} + y''_P\right) + x\left(y'_{GR} + y'_P\right) - 4\left(y_{GR} + y_P\right)$
$\left(x^2y''_{GR} + xy'_{GR} - 4y_{GR}\right) + \left(x^2y''_P + xy'_P - 4y_P\right) = 0 + x^3 = x^3$. Thus y,

having two arbitrary constants because of its term y_{GR} , is the general solution of the original equation (*).

1.3 Technology and Differential Equations

[BONUS PROBLEM/SOLUTION]

Learn the commands necessary to *solve* simple ODEs with the software you have available. Try this knowledge on the first-order nonlinear ODE $y' + y = y^3 \sin x$, which has the family of solutions $y(x) = [Ce^{2x} + \frac{2}{5}(\cos x + 2\sin x)]^{-1/2}$, where C is a constant, and $y \equiv 0$ as a singular solution. (*Verify this by hand.*) Does your computer output look like this? If not, use some algebra to get it to look like this. Do you get the singular solution from your computer? Also, see what your computer makes of the equation $|y'| + |y| = 0$.

SOLUTION. For example, in *Maple* the command

```
dsolve(diff(y(x),x)+y(x)=(y(x))^3*sin(x),y(x));
```

produces the output

$$y(x) = -\frac{\sqrt{10\cos(x) + 20\sin(x) + 25\,e^{(2x)}\ \_C1}}{2\cos(x) + 4\sin(x) + 5\,e^{(2x)}\ \_C1}$$

You should be able to see that the solution given in the exercise is the same as the one above by rationalizing the denominator of the text's solution carefully. Our CAS does not produce the singular solution.

Let's use *Maple* to solve $|y'| + |y| = 0$. We input

```
> dsolve(abs(diff(y(x),x))+abs(y(x))=0,y(x));
```

and get the output

$$y(x) = -|0|,\ y(x) = |0|,\ x - \left(\begin{cases} -\ln(y(x)) & y(x) \le 0 \\ \ln(y(x)) & 0 < y(x) \end{cases}\right) - \_C1 = 0,$$

$$x + \left(\begin{cases} -\ln(y(x)) & y(x) \le 0 \\ \ln(y(x)) & 0 < y(x) \end{cases}\right) - \_C1 = 0$$

only part of which gives us the only real-valued solution, $y \equiv 0$. You must be alert to the fact that a CAS, as in this situation, may provide complex-valued solutions to problems.

Chapter 2

First-Order Differential Equations

2.1 Separable Equations

1. $\frac{dy}{dx} = \frac{A-2y}{x} \Rightarrow \frac{dy}{A-2y} = \frac{dx}{x} \Rightarrow \int \frac{dy}{A-2y} = \int \frac{dx}{x} \Rightarrow -\frac{1}{2}\ln|A-2y| = \ln|x| + C_1 \Rightarrow$
$\ln|A-2y| = -2\ln|x| + C_2 = \ln\left(\frac{1}{x^2}\right) + C_2 \Rightarrow (\text{exponentiating})\ |A-2y| = \frac{C_3}{x^2}$, where $C_3 > 0$
$\Rightarrow A - 2y = \frac{C_4}{x^2}$, where C_4 is arbitrary $\Rightarrow y = \frac{1}{2}\left(A - \frac{C_4}{x^2}\right) = \frac{A}{2} + \frac{C}{x^2}$. Note that
$C = -C_4/2$ can be *any* real number if C_4 can be any real number.

3. $y' = 3\sqrt[3]{y^2} \Rightarrow \frac{dy}{dt} = 3\left(y^{2/3}\right) \Rightarrow \frac{dy}{y^{2/3}} = 3\,dt \Rightarrow \int y^{-2/3}\,dy = 3\int 1\,dt \Rightarrow 3y^{1/3} = 3t + C_1$
$\Rightarrow y^{1/3} = t + C_2 \Rightarrow y = (t + C_2)^3$. Now the initial condition $y(2) = 0$ implies that $0 = (0 + C_2)^3$, so that $C_2 = -2$. Therefore, $y = (t-2)^3 = t^3 - 6t^2 + 12t - 8$. But notice that in separating the variables we divided by a power of y. The solution $y \equiv 0$ is a **singular solution** of the basic ODE and satisfies the initial condition.

5. $(\cot x)y' + y = 2 \Rightarrow (\cot x)\frac{dy}{dx} = 2 - y \Rightarrow \frac{dy}{2-y} = \frac{dx}{\cot x} = \tan x\,dx \Rightarrow \int \frac{dy}{2-y} = \int \tan x\,dx$
$\Rightarrow -\ln|2-y| = -\ln|\cos x| + C_1 \Rightarrow \ln|2-y| = \ln|\cos x| + C_2 \Rightarrow |2-y| = C_3|\cos x| \Rightarrow$
$2 - y = C_4\cos x$, so that $y = 2 - C\cos x$. The initial condition implies that $-1 = y(0) = 2 - C\cos(0) = 2 - C$, so that $C = 3$ and $y = 2 - 3\cos x$. The only possible singular solution is $y \equiv 2$, but this can be obtained by letting $C = 0$.

7. $x^2y^2y' + 1 = y \Rightarrow x^2y^2\frac{dy}{dx} = y - 1 \Rightarrow \frac{y^2}{y-1}\,dy = \frac{dx}{x^2} \Rightarrow \int \frac{y^2}{y-1}\,dy = \int \frac{dx}{x^2}$
$\Rightarrow \int \left(y + 1 + \frac{1}{y-1}\right) dy = \int x^{-2}\,dx \Rightarrow \frac{y^2}{2} + y + \ln|y-1| = -\frac{1}{x} + C$. The constant

function $y \equiv 1$ is a **singular solution**. (We divided by $y - 1$ earlier. Notice that the implicit solution formula is not defined for $y = 1$.)

9. $\frac{dz}{dx} = 10^{x+z} = 10^x 10^z \Rightarrow \frac{dz}{10^z} = 10^x\, dx \Rightarrow \int \frac{dz}{10^z} = \int 10^x\, dx \Rightarrow -\frac{1}{\ln 10}10^{-z}$

$= \frac{1}{\ln 10}10^x + C_1 \Rightarrow 10^{-z} = -10^x + C \Rightarrow -z\ln 10 = \ln\left(C - 10^x\right)$

$\Rightarrow z = \frac{\ln\left(C - 10^x\right)}{\ln 10}$. Note that for each particular value of the parameter C, the solution is defined only for $10^x < C$—that is, for $x < \ln C / \ln 10$ (or $x < \log_{10} C$).

11. $(y')^2 + (x+y)y' + xy = 0 \Rightarrow (y'+x)(y'+y) = 0 \Rightarrow y' + x = 0$ or $y' + y = 0$

$\Rightarrow \frac{dy}{dx} = -x$ or $\frac{dy}{dx} = -y \Rightarrow y = -\frac{x^2}{2} + C$ or $y = Ce^{-x}$.

13. $(x+2y)\,y' = 1 \Rightarrow y' = \frac{1}{x+2y}$. Letting $z = x + 2y$, we have

$\frac{dz}{dx} = 1 + 2y' = 1 + \frac{2}{x+2y} = 1 + \frac{2}{z} = \frac{z+2}{z}$. Separating variables, we get

$\frac{z}{z+2}\,dz = dx$, or $\left(1 - \frac{2}{z+2}\right)dz = dx$, and integrating gives us

$z - 2\ln|z+2| = x + C_1$. Replacing z by $x + 2y$, we have

$x + 2y - 2\ln|x+2y+2| = x + C$; $y = -(x+2)/2$ is a singular solution.

15. $y' = \frac{x+y}{x-y} = \frac{x\left(1+\frac{y}{x}\right)}{x\left(1-\frac{y}{x}\right)} = \frac{\left(1+\frac{y}{x}\right)}{\left(1-\frac{y}{x}\right)}$. Now let $z = y/x$. As in the example,

$y' = z + x\left(\frac{dz}{dx}\right)$, so that the original equation becomes $z + x\left(\frac{dz}{dx}\right) = \frac{1+z}{1-z}$,

or $x\left(\frac{dz}{dx}\right) = \frac{1+z}{1-z} - z = \frac{1+z^2}{1-z}$. Separating variables, we get

$\left(\frac{1-z}{1+z^2}\right)dz = \left(\frac{1}{1+z^2} - \frac{z}{1+z^2}\right)dz = \frac{dx}{x}$. Integrating, we find that

$\arctan z - \frac{1}{2}\ln\left|1+z^2\right| = \ln|x| + C$. Replacing z by y/x, we get the solution

$\arctan\left(\frac{y}{x}\right) - \frac{1}{2}\ln\left(\frac{x^2+y^2}{x^2}\right) - \ln|x| - C = 0$.

17. $y' = \frac{x}{y} + \frac{y}{x}$. We have a choice here: Let $z = x/y$ or $z = y/x$. Because it will make our work a little easier, we choose $z = y/x$. Now $\frac{dy}{dx} = z + x\left(\frac{dz}{dx}\right)$ allows us to write our original equation as $\frac{dy}{dx} = z + x\left(\frac{dz}{dx}\right) = \frac{1}{z} + z$, or $x\left(\frac{dz}{dx}\right) = \frac{1}{z}$. Separating variables and integrating, we find that $\frac{z^2}{2} = \ln|x| + C_1$. Substituting y/x for z and multiplying by 2, we get $\left(\frac{y}{x}\right)^2 = 2\ln|x| + C_2$, $y^2 = 2x^2\ln|x| + C_2x^2$, so that we have two one-parameter families of solutions: $y = x\sqrt{2\ln|x| + C}$ and $y = -x\sqrt{2\ln|x| + C}$.

19. The FTC says that $\frac{d}{dx}f(x) = \frac{d}{dx}\int_0^x f(t)\,dt = f(x)$. Since $f(0) = \int_0^0 f(t)\,dt = 0$, we see that we have the IVP $y' = y$, $y(0) = 0$. Solving the equation by separating variables, we find that $y = Ce^x$. The initial condition implies that $C = 0$, so that $y = f(x) \equiv 0$.

21. (a) $\frac{dx}{dt} = x^2 \Rightarrow x^{-2}\,dx = dt \Rightarrow -x^{-1} = t + C_1 \Rightarrow \frac{1}{x} = -t + C_2 \Rightarrow x(t) = \frac{1}{C - t}$.

Now $x(1) = 1 \Rightarrow 1 = \frac{1}{C - 1} \Rightarrow C = 2 \Rightarrow x(t) = \frac{1}{2 - t}$.

(b) The interval I can be as large as $(-\infty, 2)$ or $(2, \infty)$. Any such interval I cannot include the point $t = 2$, at which $x(t)$ is not defined.

(c)

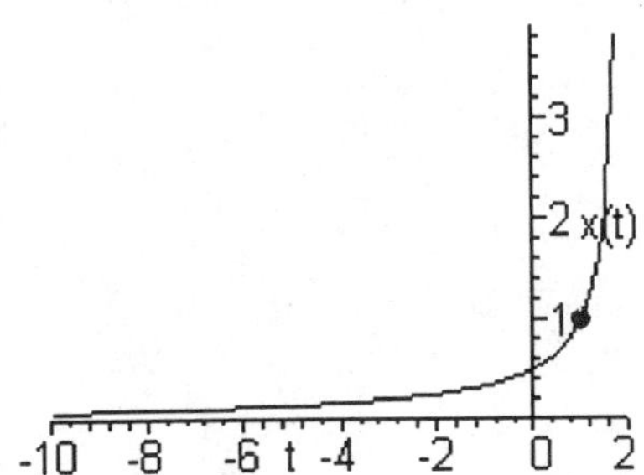

(d) Using the one-parameter formula found in (a), we want $0 = x(0) = \frac{1}{C - 0} = \frac{1}{C}$, which is impossible. However, we notice that $x \equiv 0$ is a **singular solution** that satisfies the initial condition $x(0) = 0$.

23. $\frac{dy}{dt}=-\frac{\ln 2}{30}(y-20)\Rightarrow \frac{dy}{y-20}=-\frac{\ln 2}{30}\,dt\Rightarrow \ln|y-20|=-\frac{\ln 2}{30}t+C_1\Rightarrow$

$|y-20|=C_2\,e^{-\frac{\ln 2}{30}t}=C_2\left(e^{\ln 2^{-1/30}}\right)^t=C_2\left(2^{-t/30}\right)\Rightarrow y-20=C_3\left(2^{-t/30}\right)$. Now $y(30)=60\Rightarrow 60-20=C_3\left(2^{-30/30}\right)=C_3/2,\ 40=C_3/2\Rightarrow C_3=80$. Therefore, $y=80\left(2^{-t/30}\right)+20$. Now $40=80\left(2^{-t/30}\right)+20\Rightarrow 20=80\left(2^{-t/30}\right)\Rightarrow \frac{1}{4}=2^{-t/30}$ $\Rightarrow \ln\left(\frac{1}{4}\right)=-\frac{t}{30}\ln 2\Rightarrow t=-30\frac{\ln\left(\frac{1}{4}\right)}{\ln 2}=-30(-2)=60$.

25. Separating variables and integrating, we get $\int\frac{dm}{\sqrt{1+m^2}}=\int dx=x+C$.

You can attempt to evaluate the first integral by starting with the substitution $m=\tan u$, so that $\sqrt{1+m^2}=\sec u$ and $dm=\sec^2 u\,du$—or you can consult a table of integrals to find that $\int\frac{dm}{\sqrt{1+m^2}}=\ln\left(m+\sqrt{1+m^2}\right)+K$. (We don't need an absolute value inside the logarithm because $\sqrt{1+m^2}>\sqrt{m^2}=|m|$, so that $m+\sqrt{1+m^2}>0$.) Now we have the equation $\ln\left(m+\sqrt{1+m^2}\right)=x+C$. Using the given initial condition $m(0)=0$, we find that $\ln\left(0+\sqrt{1+0^2}\right)=0+C$, so that $C=0$. Now $\ln\left(m+\sqrt{1+m^2}\right)=x\Rightarrow$ $m+\sqrt{1+m^2}=e^x$. Replacing x by $-x$, we see that $-m+\sqrt{1+m^2}=e^{-x}$, so that subtracting the second formula from the first gives us $2m=e^x-e^{-x}$, or $m=\left(e^x-e^{-x}\right)/2=\sinh(x)$, the *hyperbolic sine* of x.

27. $\frac{dL}{dt}=\frac{aL^n}{b+L^n}\Rightarrow\int\left(\frac{b+L^n}{aL^n}\right)dL=\int dt\Rightarrow\int\left(\frac{b}{a}L^{-n}+\frac{1}{a}\right)dL=\int dt\Rightarrow$

$\frac{b}{a}\left(\frac{L^{-n+1}}{-n+1}\right)+\frac{L}{a}=t+C$. Since we can assume that $L=0$ when $t=0$, we see that $C=0$. Therefore, we can write $t=\frac{1}{a}\left(L+\frac{b\,L^{1-n}}{1-n}\right)$.

29. (a) This is a separable equation: $\frac{dC}{dt}=-\frac{C}{6}\Rightarrow\int\frac{dC}{C}=-\frac{1}{6}\int dt\Rightarrow$

$\ln|C|=-\frac{t}{6}+K_1\Rightarrow|C|=K_2\,e^{\frac{-t}{6}}\Rightarrow C=K\,e^{\frac{-t}{6}}$. Since we are told that $C=14$ mg/liter at $t=0$, we can deduce that $K=14$, so that the concentration at time t can be expressed as $C = C(t) = 14e^{\frac{-t}{6}}$.

(b) The concentration becomes ineffective when $C < 5$—that is, when $14e^{\frac{-t}{6}} < 5$. Now $14e^{\frac{-t}{6}} < 5 \Rightarrow e^{\frac{-t}{6}} < 5/14 \Rightarrow -t/6 < \ln(5/14) \Rightarrow$ $t > -6\ln(5/14) \approx 6.18$ hours. Therefore a second injection should be given after about 6 hours.

(c) Mathematically, the fact that a second injection increases the concentration by 14 mg/liter means that we have a new initial condition: $C(6) = 14e^{-6/6} + 14$ $= 14e^{-1} + 14 =$ the concentration 6 hours after the first injection + the increase due to the second injection. This says that the concentration t hours after the *second* injection can be expressed as $C = (14e^{-1} + 14)e^{\frac{-t}{6}}$. Now the concentration becomes ineffective when $(14e^{-1} + 14)e^{\frac{-t}{6}} < 5$, which implies that $t > -6\ln\left(5/\left(14e^{-1} + 14\right)\right) \approx 8.06$ hours. So another injection is necessary about 8 hours after the second injection.

(d) Undesirable side effects occur when the concentration exceeds 20 mg/liter. This translates into $14e^{\frac{-t}{6}} + 14 > 20$ for the second injection. Solving the inequality gives us $t < -6\ln(3/7) \approx 5.08$ hours, which means that we should wait at least 5 hours before giving the second injection. The results of (b) and (d) say that there is an optimal "window" between 5 and 6 hours during which the first injection is still effective but a second injection won't be harmful.

(e)

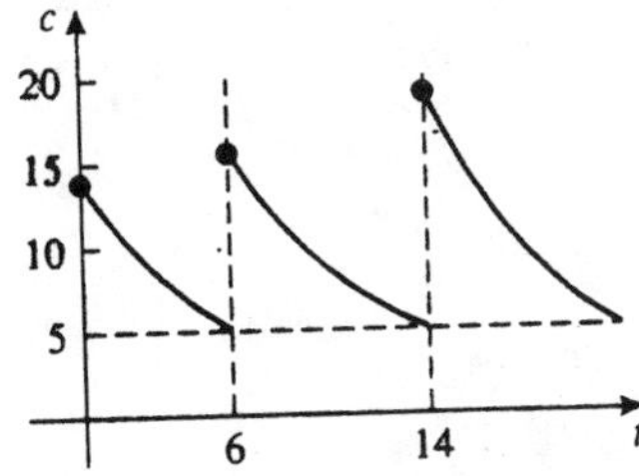

31. (a) After separating variables, we have $\int \frac{dP}{P(1-P)} = \int dt \Rightarrow \int\left(\frac{1}{1-P} + \frac{1}{P}\right) dP = \int dt$

$\Rightarrow -\ln|1-P| + \ln|P| = t + C_1 \Rightarrow \ln\left|\frac{P}{1-P}\right| = t + C_1 \Rightarrow \left|\frac{P}{1-P}\right| = C_2 e^t \Rightarrow$

$\frac{P}{1-P} = Ce^t \Rightarrow P = \frac{Ce^t}{1+Ce^t}$. You should note that $P \equiv 0$ and $P \equiv 1$ are *equilibrium solutions* (see problem 30(a)), with $P \equiv 1$ a **singular solution**.

(b) $P_0 = P(0) = C/(1+C)$ (from(a)) $\Rightarrow C = P_0/(1-P_0) \Rightarrow P(t) = \dfrac{\left(\dfrac{P_0}{1-P_0}\right)e^t}{1+\left(\dfrac{P_0}{1-P_0}\right)e^t}$

$$= \frac{P_0 e^t}{(1-P_0)+P_0 e^t} = \frac{e^t P_0}{e^t\left(P_0 + \dfrac{1-P_0}{e^t}\right)} = \frac{P_0}{\left(P_0 + \dfrac{1-P_0}{e^t}\right)} \rightarrow \frac{P_0}{(P_0+0)} = 1 \text{ as } t \to \infty.$$

(c) From the last expression in part (b), it is clear that $P(t) \to 1$ as $t \to \infty$,
whether P_0 is between 0 and 1 or is greater than 1. The only difference
between the two cases is that when $0 < P_0 < 1$,
$P(t) \to 1$ from *below* as $t \to \infty$; while for $P_0 > 1$,
$P(t) \to 1$ from *above* as $t \to \infty$. The equilibrium solution $P \equiv 1$ is a **sink**.

33 (a) We have $x(t) = \dfrac{\alpha\beta\left(1-e^{(\alpha-\beta)kt}\right)}{\beta - \alpha e^{(\alpha-\beta)kt}} = \dfrac{e^{(\alpha-\beta)kt}\left(\dfrac{\alpha\beta}{e^{(\alpha-\beta)kt}} - \alpha\beta\right)}{e^{(\alpha-\beta)kt}\left(\dfrac{\beta}{e^{(\alpha-\beta)kt}} - \alpha\right)} = \dfrac{\left(\dfrac{\alpha\beta}{e^{(\alpha-\beta)kt}} - \alpha\beta\right)}{\left(\dfrac{\beta}{e^{(\alpha-\beta)kt}} - \alpha\right)}$

$\rightarrow \dfrac{0-\alpha\beta}{0-\alpha} = \beta$. (If $\alpha > \beta$, then $\alpha - \beta > 0$ and $1/\mathrm{e}^{(\alpha-\beta)kt} \to 0$ as $t \to \infty$.)

(b) If $\alpha < \beta$, then $\alpha - \beta < 0$ and $\mathrm{e}^{(\alpha-\beta)kt} \to 0$ as $t \to \infty$. Therefore,

$$x(t) = \frac{\alpha\beta\left(1-e^{(\alpha-\beta)kt}\right)}{\beta-\alpha e^{(\alpha-\beta)kt}} \rightarrow \frac{\alpha\beta(1-0)}{\beta-\alpha\cdot 0} = \frac{\alpha\beta}{\beta} = \alpha \quad \text{as } t \to \infty.$$

2.2 Linear Equations

1. $y' + 2y = 4x \Rightarrow$ integrating factor $\mu(x) = e^{\int 2\,dx} = e^{2x}$. Then $e^{2x}(y' + 2y) = e^{2x} \cdot 4x$

$\Rightarrow e^{2x}y' + 2e^{2x}y = e^{2x}4x \Rightarrow \frac{d}{dx}\left[e^{2x}y\right] = 4xe^{2x} \Rightarrow e^{2x}y = (2x-1)e^{2x} + C \Rightarrow$

$y = 2x - 1 + Ce^{-2x}$. Note that the solution curves are asymptotic to the line $y = 2x - 1$ as t tends to infinity.

3. $\dot{x} + 2tx = t^3 \Rightarrow \mu(t) = e^{\int 2t\,dt} = e^{t^2} \Rightarrow \frac{d}{dt}\left[e^{t^2}x\right] = t^3 e^{t^2} \Rightarrow e^{t^2}x = \int t^3 e^{t^2}\,dt$

$= \left(t^2 - 1\right)\frac{e^{t^2}}{2} + C \Rightarrow x = \frac{t^2}{2} - \frac{1}{2} + Ce^{-t^2}$.

5. $t\,y' = -3y + t^3 - t^2 \Rightarrow t\,y' + 3y = t^3 - t^2 \Rightarrow y' + \left(\frac{3}{t}\right)y = t^2 - t \Rightarrow \mu(t) = e^{\int \frac{3}{t}\,dt} = e^{3\ln t}$

$= t^3 \Rightarrow \frac{d}{dt}\left[t^3 y\right] = t^5 - t^4 \Rightarrow t^3 y = \frac{t^6}{6} - \frac{t^5}{5} + C \Rightarrow y = \frac{t^3}{6} - \frac{t^2}{5} + \frac{C}{t^3}$.

7. $y = x(y' - x\cos x) = x\,y' - x^2\cos x \Rightarrow x\,y' - y = x^2\cos x \Rightarrow y' + \left(\frac{-1}{x}\right)y = x\cos x$

$\Rightarrow \mu(x) = e^{\int -\frac{1}{x}\,dx} = e^{-\ln x} = 1/x \Rightarrow \frac{d}{dx}\left[\frac{y}{x}\right] = \cos x \Rightarrow \frac{y}{x} = \int \cos x\,dx = \sin x + C$

$\Rightarrow y = x\sin x + Cx$.

9. $t(x' - x) = (1 + t^2)e^t \Rightarrow x' - x = \left(\frac{1+t^2}{t}\right)e^t \Rightarrow \mu(t) = e^{\int -1\,dt} = e^{-t} \Rightarrow \frac{d}{dt}\left[e^{-t}x\right]$

$= \left(\frac{1+t^2}{t}\right) = \frac{1}{t} + t \Rightarrow e^{-t}x = \int\left(\frac{1}{t} + t\right)dt = \ln|t| + t^2/2 + C \Rightarrow$

11. $x\,y' + y - e^x = 0 \Rightarrow x\,y' + y = e^x \Rightarrow y' + \left(\frac{1}{x}\right)y = \frac{e^x}{x} \Rightarrow \mu(x) = e^{\int \frac{1}{x}\,dx} = x$

$\Rightarrow \frac{d}{dx}[x\,y] = e^x \Rightarrow x\,y = \int e^x\,dx = e^x + C \Rightarrow y = \frac{e^x + C}{x}$. The initial condition

$y(a) = b$ implies that $b = \frac{e^a + C}{a}$, so that $C = ab - e^a$ and we have

$y(x) = \frac{e^x + ab - e^a}{x}$. (In the initial condition, clearly a shouldn't be 0.)

13. $y' + ay = e^{mx} \Rightarrow \mu(x) = e^{\int a\,dx} = e^{ax} \Rightarrow \frac{d}{dx}\left[e^{ax}y\right] = e^{ax}e^{mx} = e^{(a+m)x}$

$e^{ax}y=\int e^{(a+m)x}\,dx=\dfrac{e^{(a+m)x}}{a+m}+C$, if $a+m\neq 0$—that is, if $m\neq -a$. For $m\neq -a$, we have $y=\dfrac{e^{mx}}{a+m}+Ce^{-ax}$. If $m=-a$, then

$e^{ax}y=\int e^{(a-a)x}\,dx=\int 1\,dx=x+C$, so that $y=xe^{-ax}+Ce^{-ax}=(x+C)\,e^{-ax}$.

Note: A CAS that can solve ODEs may miss the need for an analysis of two cases.

15. $tx'-\left(\dfrac{x}{t+1}\right)=t\Rightarrow x'+\left(\dfrac{-1}{t(t+1)}\right)x=1\Rightarrow \mu(t)=e^{\int\frac{-1}{t(t+1)}dt}=e^{\int\left(\frac{1}{t+1}-\frac{1}{t}\right)dt}$

$=e^{\ln|t+1|-\ln|t|}=e^{\ln\left|\frac{t+1}{t}\right|}=\dfrac{t+1}{t}\Rightarrow\dfrac{d}{dt}\left[\dfrac{t+1}{t}x\right]=\dfrac{t+1}{t}\Rightarrow\dfrac{t+1}{t}x=\int\dfrac{t+1}{t}\,dt$

$=t+\ln|t|+C\Rightarrow x=\left(\dfrac{t}{t+1}\right)\cdot\left(t+\ln|t|+C\right)$. Now $x(1)=0$ implies that

$0=\left(\dfrac{1}{1+1}\right)\left(1+\ln 1+C\right)$, so that C = -1 and $x(t)=\left(\dfrac{t}{t+1}\right)\left(t+\ln|t|-1\right)$.

17. $x\left(e^{y}-y'\right)=2$: Unlike the situation in Exercise 16, just switching the roles of y and x doesn't work. The resulting equation would not be linear in either x or y. Noticing that $\left(e^{-y}\right)'=-y'e^{-y}$, we can multiply each side of the differential equation by e^{-y} to get $x\left(1-e^{-y}y'\right)=2e^{-y}$. Making the substitution $z=e^{-y}$ gives us $x\left(1+\dfrac{dz}{dx}\right)=2z$, $x+x\dfrac{dz}{dx}=2z$, $x\dfrac{dz}{dx}-2z=-x$, and $\dfrac{dz}{dx}+\left(\dfrac{-2}{x}\right)z=-1$, *a linear equation in* z. An integrating factor is $e^{\int -\frac{2}{x}dx}=\dfrac{1}{x^2}$. Therefore we have

$\dfrac{d}{dx}\left[\dfrac{z}{x^2}\right]=-\dfrac{1}{x^2}\Rightarrow\dfrac{z}{x^2}=\int -\dfrac{1}{x^2}\,dx=\dfrac{1}{x}+C\Rightarrow z=x+Cx^2$, or

$e^{-y}=x+Cx^2\Rightarrow y=-\ln\left(x+Cx^2\right)$.

19. $y'=\dfrac{4}{t}y-6ty^2$, or $y'+\left(-\dfrac{4}{t}\right)y=-6ty^2$: Here $n=2$. Divide both sides by y^2 to get $y^{-2}y'+\left(\dfrac{-4}{t}\right)y^{-1}=-6t$. Letting $z=y^{1-2}=y^{-1}$, we have $\dfrac{dz}{dt}=-y^{-2}\cdot\dfrac{dy}{dt}$, so that the equation becomes $-\dfrac{dz}{dt}+\left(-\dfrac{4}{t}\right)z=-6t$, or $\dfrac{dz}{dt}+\left(\dfrac{4}{t}\right)z=6t$, linear in z.

Now $\mu(t)=e^{\int\frac{4}{t}dt}=t^4$, so that we have $\dfrac{d}{dt}\left[t^4z\right]=6t^5$, $t^4z=\int 6t^5\,dt=t^6+C$,

$z = \dfrac{t^6 + C}{t^4}$, $y = \dfrac{1}{z} = \dfrac{t^4}{t^6 + C}$. Note that $y \equiv 0$ is a **singular solution**.

21. $\dfrac{dy}{dx} + y = xy^3$: Here $n = 3$, so that we let $z = y^{1-3} = y^{-2}$ and $\dfrac{dz}{dx} = -2y^{-3}\dfrac{dy}{dx}$.

Dividing both sides of the original equation by y^3, we get $y^{-3}\dfrac{dy}{dx} + y^{-2} = x$,

or $-\frac{1}{2}\dfrac{dz}{dx} + z = x$, $\dfrac{dz}{dx} - 2z = -2x$, so that $\mu(x) = e^{\int -2\,dx} = e^{-2x}$. Then

$\dfrac{d}{dx}\left[e^{-2x}z\right] = -2xe^{-2x}$, or $e^{-2x}z = \int -2xe^{-2x}\,dx = \frac{1}{2}e^{-2x}(2x+1) + C$, so

that $z = \frac{1}{2}(2x+1) + Ce^{2x} = x + \dfrac{1}{2} + Ce^{2x}$ and $\dfrac{1}{y^2} = x + \dfrac{1}{2} + Ce^{2x}$, or

$y = \dfrac{\pm 1}{\sqrt{x + \frac{1}{2} + Ce^{2x}}}$. Note that $y \equiv 0$ is a **singular solution**.

23. (a) $\dfrac{dW}{dt} = \alpha W^{2/3} - \beta W$, $\dfrac{dW}{dt} + \beta W = \alpha W^{2/3}$: Here $n = 2/3$. Letting $z = W^{1-2/3}$

$= W^{1/3}$ and dividing both sides of the ODE by $W^{2/3}$ gives us the new equation

$W^{-2/3}\dfrac{dW}{dt} + \beta W^{1/3} = \alpha$. Noting that $\dfrac{dz}{dt} = \frac{1}{3}W^{-2/3}\cdot\dfrac{dW}{dt}$, we can write this last

equation as $3\dfrac{dz}{dt} + \beta z = \alpha$, or $\dfrac{dz}{dt} + \left(\dfrac{\beta}{3}\right)z = \dfrac{\alpha}{3}$, so that $\mu(t) = e^{\int \frac{\beta}{3}\,dt} = e^{\beta t/3}$.

Now $\dfrac{d}{dt}\left[e^{\beta t/3}z\right] = \dfrac{\alpha}{3}e^{\beta t/3} \Rightarrow e^{\beta t/3}z = \int \dfrac{\alpha}{3}e^{\beta t/3}\,dt = \dfrac{\alpha}{\beta}e^{\beta t/3} + C \Rightarrow$

$z = \dfrac{\alpha}{\beta} + Ce^{-\beta t/3}$. Replacing z by $W^{1/3}$, we find that $W^{1/3} = \dfrac{\alpha}{\beta} + Ce^{-\beta t/3}$,

or $W(t) = \left(\dfrac{\alpha}{\beta} + Ce^{-\beta t/3}\right)^3$.

(b) $W_\infty = \lim_{t\to\infty} W(t) = \lim_{t\to\infty}\left(\dfrac{\alpha}{\beta} + Ce^{-\beta t/3}\right)^3 = \left(\dfrac{\alpha}{\beta} + C\cdot\lim_{t\to\infty} e^{-\beta t/3}\right)^3 = \left(\dfrac{\alpha}{\beta}\right)^3$.

(c) $W(0) = 0 \Rightarrow 0 = W(0) = \left(\dfrac{\alpha}{\beta} + Ce^0\right)^3 = \left(\dfrac{\alpha}{\beta} + C\right)^3 \Rightarrow C = -\dfrac{\alpha}{\beta}$. Therefore

$W(t) = \left(\dfrac{\alpha}{\beta} - \dfrac{\alpha}{\beta}e^{-\beta t/3}\right)^3 = \left(\dfrac{\alpha}{\beta}\right)^3\left(1 - e^{-\beta t/3}\right)^3 = W_\infty\left(1 - e^{-\beta t/3}\right)^3$.

(d)

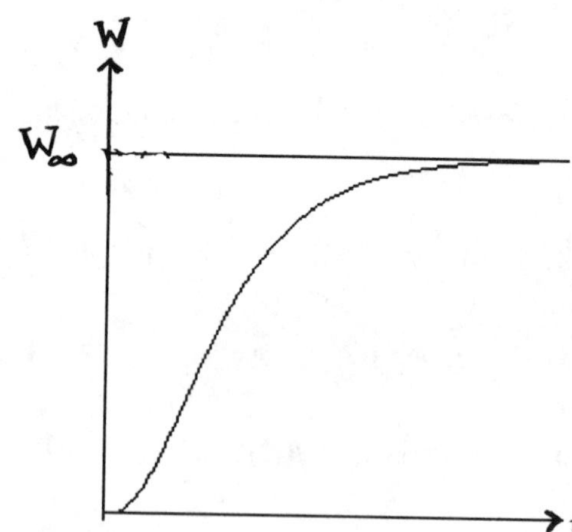

25. (a) $L\frac{dI}{dt}+RI=E \Rightarrow \frac{dI}{dt}+\left(\frac{R}{L}\right)I=\frac{E}{L} \Rightarrow \mu(t)=e^{\int \frac{R}{L}dt}=e^{(R/L)t} \Rightarrow \frac{d}{dt}\left[e^{(R/L)t}I\right]$

$=\frac{E}{L}e^{(R/L)t} \Rightarrow e^{(R/L)t}I=\int \frac{E}{L}e^{(R/L)t}dt=\frac{E}{R}e^{(R/L)t}+C \Rightarrow I=\frac{E}{R}+Ce^{-(R/L)t}.$

Now the wording of the problem suggests that $I(0)=0$, so that

$0=I(0)=\frac{E}{R}+Ce^{0} \Rightarrow C=-\frac{E}{R}$. Therefore,

$I(t)=\frac{E}{R}-\frac{E}{R}e^{-(R/L)t}=\frac{E}{R}\left(1-e^{-(R/L)t}\right).$

(b) $\lim_{t\to\infty} I(t)=\lim_{t\to\infty}\frac{E}{R}\left(1-e^{-(R/L)t}\right)=\frac{E}{R}\left(1-\lim_{t\to\infty}e^{-(R/L)t}\right)=\frac{E}{R}.$

(c) By "final " value, we mean the value E/R found in (b). Now we want $I=\frac{1}{2}\frac{E}{R}$, or $\frac{1}{2}\frac{E}{R}=\frac{E}{R}\left(1-e^{-(R/L)t}\right)$, so that $\frac{1}{2}=1-e^{-(R/L)t}, e^{-(R/L)t}=\frac{1}{2}$, $-\frac{R}{L}t=\ln\left(\frac{1}{2}\right)=-\ln 2$, $\frac{R}{L}t=\ln 2$, and $t=\frac{L}{R}\ln 2$.

(d) Using the general solution found in part (a), we see that $I(0)=\frac{E}{R}$ $\Rightarrow \frac{E}{R}=\frac{E}{R}\left(1-Ce^{0}\right)=\frac{E}{R}+C^*$, implying that $C^*=0$. Therefore $I(t)\equiv\frac{E}{R}$, which we realize is the *equilibrium solution* (see problem 30(a) in Exercises 2.1) of the autonomous equation.

27. The equation is $R\frac{dQ}{dt}+\frac{Q}{C}=E_0\sin(\omega t)$, with $Q(0)=0$. Then

$$\frac{dQ}{dt}+\left(\frac{1}{RC}\right)Q=\frac{E_0}{R}\sin(\omega t)\Rightarrow\mu(t)=e^{\int\frac{1}{RC}dt}=e^{t/RC}\Rightarrow\frac{d}{dt}\left[e^{t/RC}Q\right]$$

$$=\frac{E_0}{R}e^{t/RC}\sin(\omega t)\Rightarrow e^{t/RC}Q=\frac{E_0}{R}\int e^{t/RC}\sin(\omega t)\,dt$$

$$=\frac{E_0}{R}\left\{\frac{RCe^{t/RC}\left[\sin(\omega t)\ -\ \omega RC\cos(\omega t)\right]}{1+(RC\omega)^2}\right\}+K,\ \text{so that}$$

$$Q(t)=\left\{\frac{E_0C\left[\sin(\omega t)\ -\ \omega RC\cos(\omega t)\right]}{1+(RC\omega)^2}\right\}+Ke^{-t/RC}.\ \text{Now}\ 0=Q(0)$$

$$=\frac{E_0C\left[0-\omega RC\right]}{1+(RC\omega)^2}+K=-\frac{\omega E_0RC^2}{1+(RC\omega)^2}+K,\ \text{so that}\ K=\frac{\omega E_0RC^2}{1+(RC\omega)^2}\ \text{and}$$

$$Q(t)=\frac{E_0C\left[\sin(\omega t)\ -\ \omega RC\cos(\omega t)\right]}{1+(RC\omega)^2}+\frac{\omega E_0RC^2}{1+(RC\omega)^2}e^{-t/RC}$$

$$=\frac{E_0C}{1+(RC\omega)^2}\left\{\sin(\omega t)\ -\ \omega RC\cos(\omega t)+\omega RCe^{-t/RC}\right\}.$$

29. For $0<t<T$, the equation is $\frac{dS}{dt}+\left(\frac{r\bar{A}}{M}+\lambda\right)S=r\bar{A}$. Let $b=\frac{r\bar{A}}{M}+\lambda$. Then

$\mu(t)=e^{\int b\,dt}=e^{bt}$ and we have $\frac{d}{dt}\left[e^{bt}S\right]=e^{bt}r\bar{A}\Rightarrow e^{bt}S=r\bar{A}\int e^{bt}\,dt$

$=\frac{r\bar{A}}{b}e^{bt}+C\Rightarrow S(t)=\frac{r\bar{A}}{b}+Ce^{-bt}$ for $0<t<T$. Letting $S_0=S(0)$, we get

$S_0=\frac{r\bar{A}}{b}+C$, so that $C=S_0-\frac{r\bar{A}}{b}$ and $S(t)=\frac{r\bar{A}}{b}+\left(S_0-\frac{r\bar{A}}{b}\right)e^{-bt}$ (*)

for $0<t<T$. Now for $t>T$, $A=0$ and the equation becomes $\frac{dS}{dt}+\lambda S=0$,

or $\frac{dS}{dt}=-\lambda S$, which is separable and has solution $S=ke^{-\lambda t}$. At $t=T$,

let $S=S_T$, so that $S_T=ke^{-\lambda t}$ (**). Hence for $t\geq T$, $S(t)=S_Te^{-\lambda(t-T)}$. ,

From (*)we see that S_T has the value $\frac{r\bar{A}}{b}+\left(S_0-\frac{r\bar{A}}{b}\right)e^{-bT}$. Combining

(*) and (**) and substituting for b, we get the formula for predicted sales:

$$S(t)=\begin{cases}\dfrac{r\bar{A}}{\left(\frac{r\bar{A}}{M}+\lambda\right)}+\left(S_0-\dfrac{r\bar{A}}{\left(\frac{r\bar{A}}{M}+\lambda\right)}\right)e^{-\left(\frac{r\bar{A}}{M}+\lambda\right)t} & \text{for } 0<t<T\\ S_Te^{-\lambda(t-T)} & \text{for } t\geq T\end{cases}$$

(b) Choosing $\overline{A} = 1000$, $r = 10$, $\lambda = 0.1$, $S_0 = 20000$, $S_T = 36000$, $M = 60000$, and $T = 10$, we have the following graph:

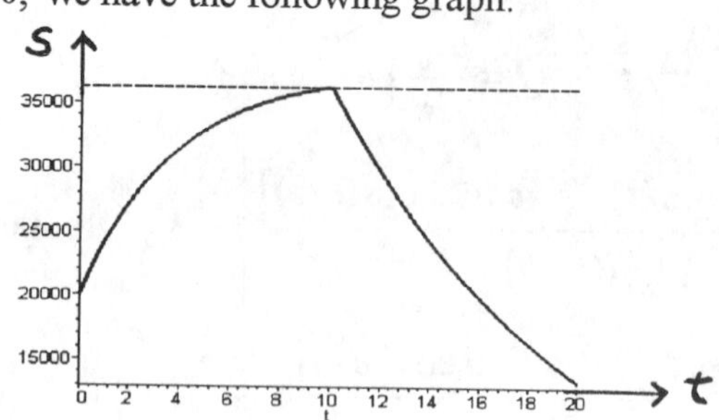

31. First note that the volume of material in the tank changes with time. Then **net rate = rate of inflow - rate of outflow**, so that the volume of mixture in the tank is increasing at the net rate of 4 gal/min - 3 gal/min = 1 gal/min. This says that $dV/dt = 1$, so that $V(t) = t + C$. Since $V(0) = 50$ gal, we have $V(t) = t + 50$. If we let $Q(t)$ denote the *quantity* of potassium (in grams) in the tank at time t, with $Q(0) = 0$, then the *concentration* in the tank at time t is $Q / (t + 50)$, in units of gm/gal. Therefore we have

$$\underbrace{\frac{dQ}{dt}}_{\substack{\text{net rate of change of} \\ \text{potassium in tank (gm/min)}}} = \underbrace{\left(4\,\frac{gal}{\min}\cdot 10\,\frac{gm}{gal}\right)}_{\substack{\text{rate of inflow} \\ \text{(gm/min)}}} - \underbrace{\left(3\,\frac{gal}{\min}\cdot\frac{Q(t)}{t+50}\,\frac{gm}{gal}\right)}_{\substack{\text{rate of outflow} \\ \text{(gm/min)}}},$$

so that we have $\dfrac{dQ}{dt} + \left(\dfrac{3}{t+50}\right)Q = 40$, a linear equation with integrating factor $\mu(t) = e^{\int \frac{3}{t+50}dt} = (t+50)^3$. Then $\dfrac{d}{dt}\left[(t+50)^3 Q\right] = 40(t+50)^3$, so that $(t+50)^3 Q = \int 40(t+50)^3\,dt = 10(t+50)^4 + C$, or $Q(t) = \dfrac{10(t+50)^4 + C}{(t+50)^3}$

$= 10(t+50) + \dfrac{C}{(t+50)^3}$. The initial condition Q(0) = 0 $\Rightarrow 0 = 10(50) + \dfrac{C}{(50)^3}$, so that $C = -10(50)^4$ and $Q(t) = 10(t+50) - \dfrac{10(50)^4}{(t+50)^3}$. Since $V(t) = t + 50$ and the tank holds 100 gallons, the tank will be full when $t = 50$. Then $Q(50)$

$= 10(50+50) - \dfrac{10(50)^4}{(50+50)^3} = 937.5$ gms and the concentration of potassium at this time is $\dfrac{937.5 \text{ gms}}{100 \text{ gal}} = 9.375$ gms / gal.

33. (a) If $X(t)$ is the amount (in grams, for example) of chlorine in solution at time t, then the rate at which chlorine is entering the tank is (0.01 gm/gal)(2 gal/sec) = 2/100 gm/sec and the rate at which chlorine is leaving is

$$\left(\frac{X}{200-t}\,\frac{\text{gm}}{\text{gal}}\right)\cdot\left(3\,\frac{\text{gal}}{\text{sec}}\right) = \frac{3X}{200-t}\,\frac{\text{gm}}{\text{sec}}.$$

Note that the net amount of liquid in the tank is changing at the rate 2 gal/sec - 3 gal/sec = -1 gal/sec, so that the amount of liquid in the tank at any time t is given by 200 - t. Using the principle **net rate = rate of inflow – rate of outflow**, we get the equation

$$\frac{dX}{dt} = \frac{2}{100} - \frac{3X}{200-t}, \text{ or } \frac{dX}{dt} + \left(\frac{3}{200-t}\right)X = \frac{2}{100}.$$

Multiplying by the integrating factor $e^{\int \frac{3}{200-t}\,dt} = e^{-3\ln(200-t)} = (200-t)^{-3}$ and integrating, we get

$$X(t) = \tfrac{1}{100}(200-t) + C(200-t)^3.$$

Since $X(0) = 0$, we find that $0 = \frac{200}{100} + C(200)^3$, or $C = -\frac{200}{100(200)^3} = -\frac{2}{(200)^3}$.

Therefore $X(t) = \frac{1}{100}(200-t) - \frac{2}{(200)^3}(200-t)^3$. Now the tank is half full when 200 - t = 100, or t = 100, so that the concentration of chlorine at this time is $\dfrac{X(100)}{100} = \dfrac{100}{10000} - \dfrac{2(100)^3}{100(200)^3} = 0.0075 = 0.75$ % solution.

(b) When the tank is half full, it contains 100 gallons. If 100 gallons of 1% solution is added, then 1 gram (= 100 gal × 0.01 gm/gal) is added to the ¾ gram (= 100 gal × 0.0075 gm/gal), making a total of 1.75 grams of chlorine in the 200 gallon tank, resulting in a concentration of 1.75/200 = 0.00875 gm/gal = 0.875 %.

35. (a) This is a compartment problem, with the agency as the compartment. Let $W(t)$ be the *number* of women at time t, with $W(0)$ = 25% of 6000 = 1500. Now the net change in total staff is 50 - 100 = -50 people/week, so that *the staff size at time* t *is* 6000 - 50 t *people*. Summarizing this information, we have

$$\underbrace{\frac{dW}{dt}}_{\substack{\text{rate of change}\\ \text{in no. of women}}} = \underbrace{25 \text{ women / week}}_{\substack{\text{rate of inflow of women}\\ = 50\% \text{ of all replacements}}} - \underbrace{\left(\underbrace{100 \text{ people / week}}_{\text{rate of people leaving}} \cdot \underbrace{\frac{W(t)}{6000-50t} \text{ women / people}}_{\substack{\text{proportion of women on staff}\\ \text{at time of leaving}}} \right)}_{\text{rate of women leaving}}$$

or, $\frac{dW}{dt} + \left(\frac{100}{6000-50t}\right)W = 25$. The integrating factor is

$\mu(t) = (6000-50t)^{-2}$, so that we get $W(t) = \frac{1}{2}(6000-50t) + C(6000-50t)^2$. Since $W(0) = 1500$, we find that $C = -1/24000$. Thus $W(t) = \frac{1}{2}(6000-50t) - \frac{1}{24000}(6000-50t)^2$, and when $t = 40$ we have the staff total equal to 6000 - 50(40) = **4000** and $W(40) =$ 2000 - 2000/3, so that **the staff is about (2000 - 2000/3) / 4000 = 1/3 or $33\frac{1}{3}$ % female**.

(b) If *all* new employees were women, the differential equation would be $\frac{dW}{dt} = 50 - \left(\frac{100}{6000-50t}\right)W$, which leads to

$W(t) = (6000-50t) + C(6000-50t)^2$. Now $W(t) = 1500$ when $t = 0$, so that $C = -1/8000$. Then when $t = 40$, $W(t) = 2000$, which is one-half, or **50%**, of the staff at that time.

37. (a) If $Q(x) \equiv 0$, then the equation has the form $\frac{dy}{dx} + P(x)y = 0$, or $\frac{dy}{dx} = -P(x)y$,

so that we can separate the variables and get $\frac{dy}{y} = -P(x)\,dx$. Integrating, we get

$\ln|y| = -\int P(x)dx + C_1$, so that $|y| = C_2 e^{-\int P(x)dx}$ and $y = Ce^{-\int P(x)dx}$. This is the general solution, y_{GH}, of the homogeneous equation.

(b) Letting $y(x) = e^{-\int P(x)dx} \cdot \int e^{\int P(x)dx} Q(x)dx$, we see (Product Rule and FTC) that

$dy/dx = e^{-\int P(x)dx} \cdot e^{\int P(x)dx} Q(x) - P(x)e^{-\int P(x)dx} \cdot \int e^{\int P(x)dx} Q(x)dx =$

$Q(x) - P(x)e^{-\int P(x)dx} \cdot \int e^{\int P(x)dx} Q(x)dx$ and $dy/dx + P(x)y =$

$Q(x) - P(x)e^{-\int P(x)dx} \cdot \int e^{\int P(x)dx} Q(x)dx + P(x) \cdot \left(e^{-\int P(x)dx} \int e^{\int P(x)dx} Q(x)dx\right)$

$= Q(x)$, so that $y(x)$ is a particular solution of the nonhomogeneous equation. Thus formula (2.6.2) can be expressed as $y_{GNH} = y_{GH} + y_{PNH}$.

(c) Since $L(y)=\frac{dy}{dx}+P(x)y$ is a linear operator, the Superposition Principle yields $L(y_{\text{GNH}})=L(y_{\text{GH}}+y_{\text{PNH}})=L(y_{\text{GH}})+L(y_{\text{PNH}})=0+Q(x)=Q(x)$, as expected.

2.3 Slope Fields

1.

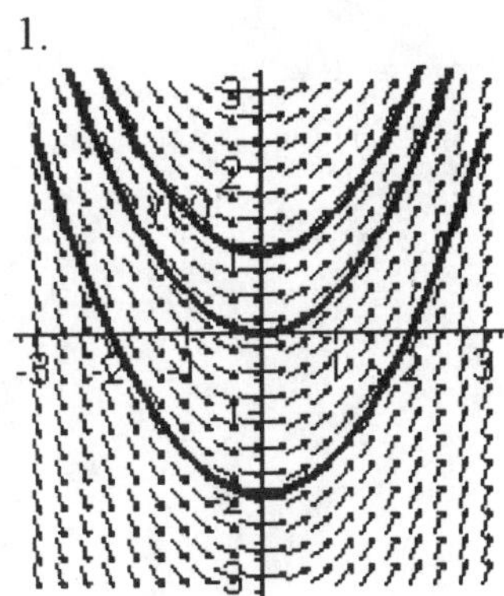

3.

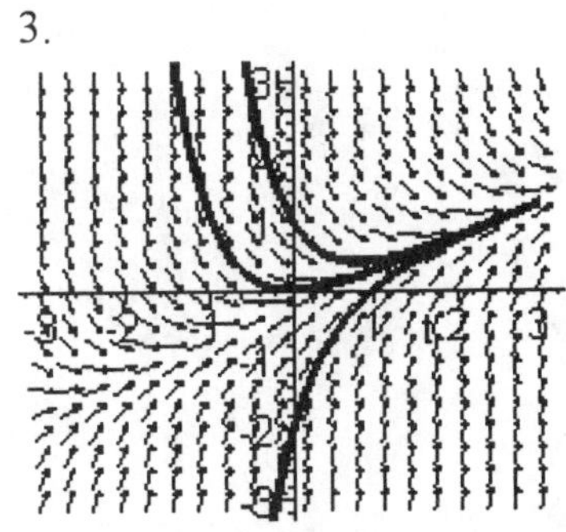

5.

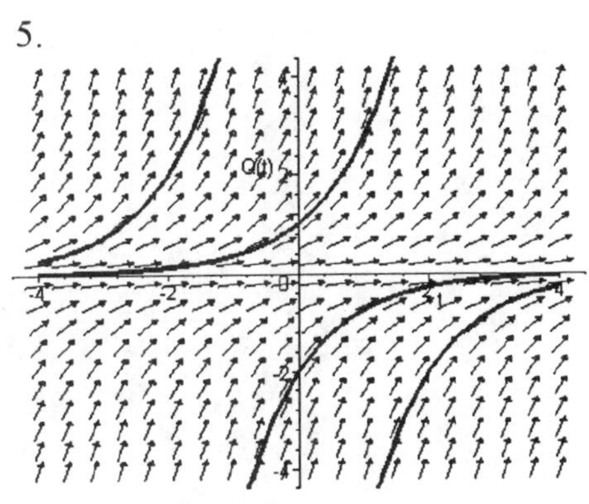

7.

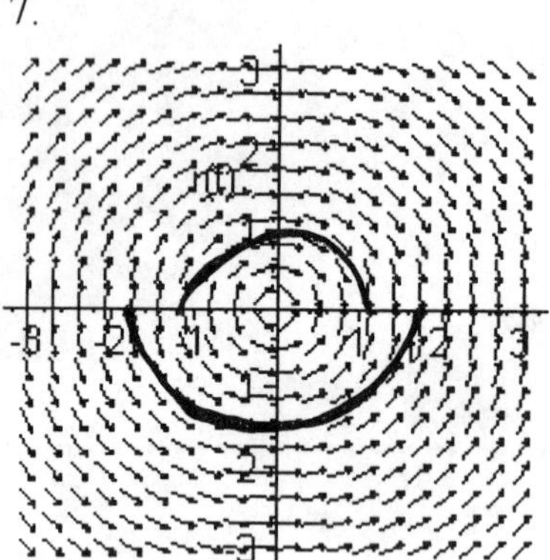

9.

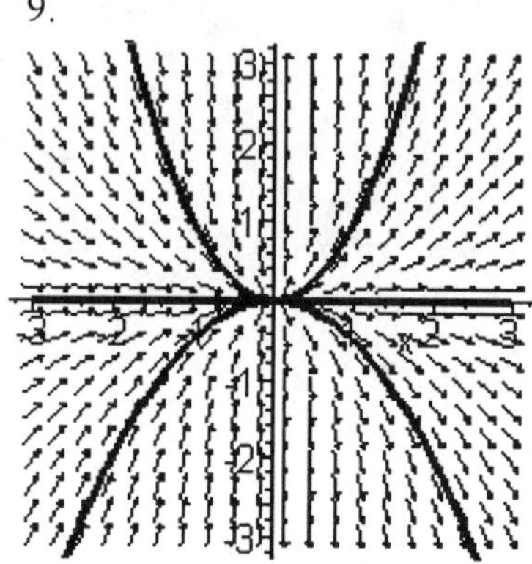

11.

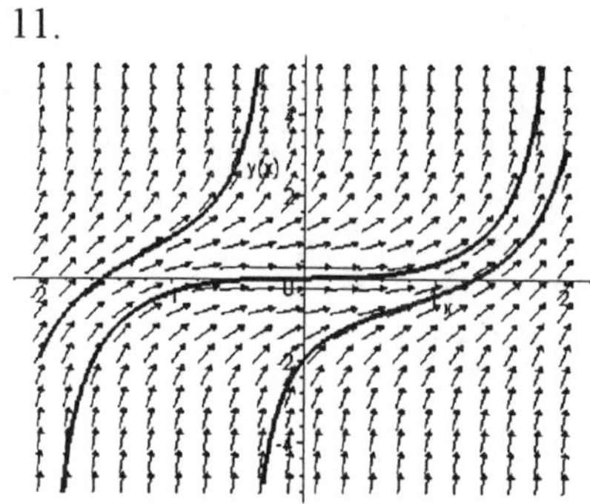

13.

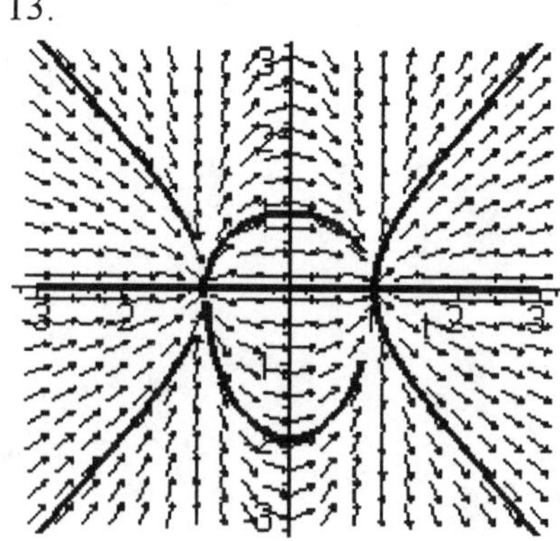

15.

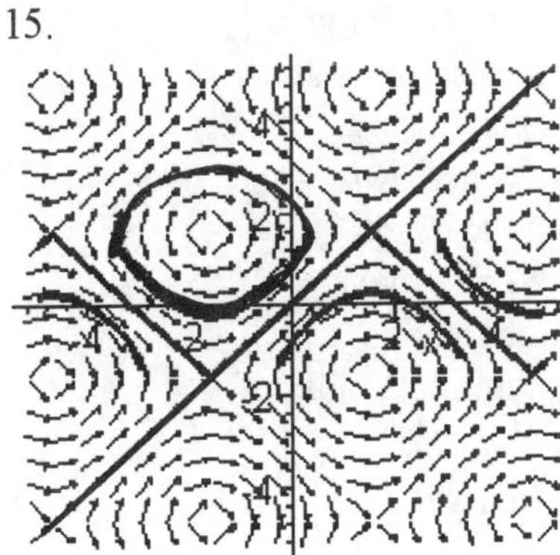

17.

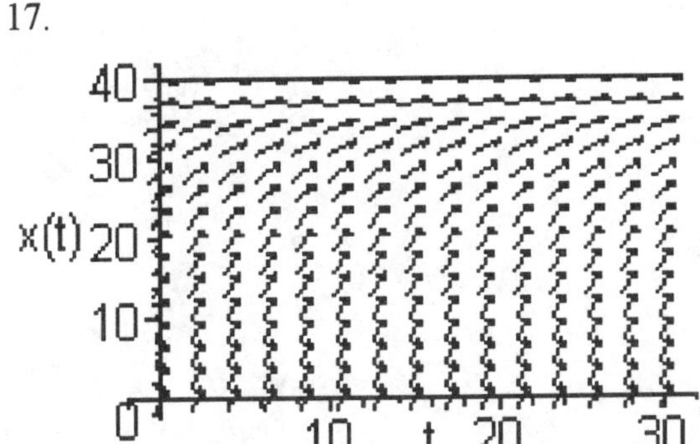

19. The isoclines are the curves defined by $dy/dt = C$, where C is a constant. In this case we have $(y+t)/(y-t) = C$, $y + t = C(y - t) = Cy - Ct$, so that $y - Cy = -t - Ct$, or $y = -((1+C)/(1-C))t$. For $C \neq 1$, this describes a one-parameter family of straight lines through the origin. For $C = 1$, we get the y-axis as the isocline.

21. Suppose that $x = \varphi(t)$ is a solution of the autonomous equation $dx/dt = f(x)$. If k is any real number, then $[\varphi(t+k)]' = \varphi'(t+k)\cdot(t+k)' = f'(t+k) = f(x)$.

23. Equation (a) is *nonautonomous*. Therefore along any horizontal line $y = k$ in the slope field, the slopes will change as the value of t changes. Equation (b) is also *nonautonomous*, but the slope field depends only on the value of the independent variable t. Here every solution curve has the form $y = \int_{t_0}^{t} f(x)\,dx + y_0$. Equation (c) is *autonomous*, so that along any horizontal line $y = k$ in the slope field, the slopes will be constant.

25. Equation (1) is autonomous and can only match slope field (C) or (D). Noting that $dy/dt = 0$ only for $y = -1$, we conclude that (1) matches (C). We can also see that $dy/dt < 0$ for $y < -1$, again giving us (C) as the match. Equation (2) is nonautonomous, giving us (A) or (B) as the only possible matches. Since $dy/dt = 0$ only for $y = t$, we look for horizontal "steps" along this line through the origin. Slope field (A) has this feature. We also note that $dy/dt > 0$ for $y > t$ and $dy/dt < 0$ for $y < t$, a feature present in slope field (A). Equation (3) is nonautonomous, with the vertical line $t = -1$ as its only nullcline. Furthermore, by integrating both sides of the equation with respect to t, we find that the solutions are the parabolas $y = t^2/2 + t + C$. Only slope field (B) has the two features described.

27.

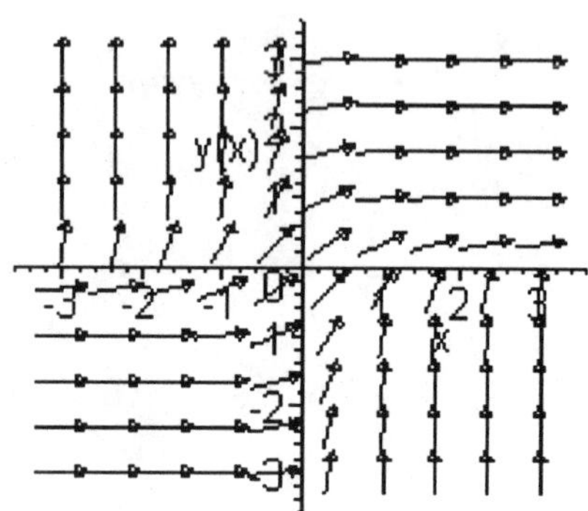

The slope field indicates that any solution must be an increasing function. Analytically, the fact that exp(-2 x y) is always positive tells us this. Some solutions in the second quadrant seem to have vertical asymptotes, so that they "blow up in finite time," while other solutions starting out in this area flatten out (approach some finite value asymptotically) as they pass through the first quadrant. Solutions with initial points in the third quadrant are almost flat until they pass into the first or fourth quadrants. Starting out in the fourth quadrant, a solution will start out having a very large slope, but will move into the first quadrant and approach a positive finite value asymptotically. Overall then, we see that as $x \to \infty$, we have both $y \to \infty$ and $y \to a$, where a is a positive real number. As $x \to -\infty$ (i.e., as we look at the slope field from right to left), we see that $y \to \infty$ or $y \to 0$.

2.4 Phase Lines and Phase Portraits

<It's useful to look at the slope fields for the equations in Exercises 1 - 12 as a check.>

1. We have $dy/dt = y^2 - 1 = (y + 1)(y - 1)$, so that the critical points are $y = -1$ and $y = 1$. These points split the y-axis into three intervals: $-\infty < y < -1$, $-1 < y < 1$, and $1 < y < \infty$. If $y < -1$, then $dy/dt > 0$. If $-1 < y < 1$, then $dy/dt < 0$. Finally, for $y > 1$, $dy/dt > 0$. The resulting phase portrait is

3. The critical points are $x = -1$ and $x = 3$. In the subinterval $x < -1$, $x' > 0$. In the subinterval $-1 < x < 3$, $x' < 0$. For $x > 3$, we have $x' > 0$:

5. We have critical points where $e^y = 1$—that is, when $y = 0$. For $y < 0$, $y' < 0$; while for $y > 0$, we have $y' > 0$:

7. This is similar to Exercise 4. The critical points are $y = k\pi$, where $k = 0, \pm 1, \pm 2, \ldots$. Examination of the graph of the sine reveals that $y' > 0$ on the intervals $(2k\pi, (2k+1)\pi)$, $k = 0, \pm 1, \pm 2, \ldots$ and $y' < 0$ on the intervals $((2k-1)\pi, 2k\pi)$, $k = 0, \pm 1, \pm 2, \ldots$:

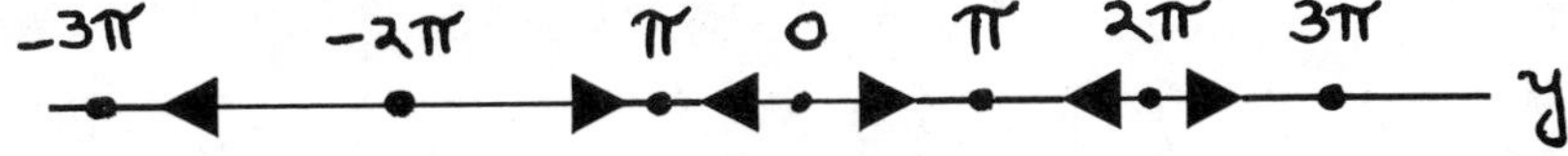

9. The only critical point is $y = 0$. It is easy to see that $y' > 0$ when $y > 0$ and $y' < 0$ when $y < 0$:

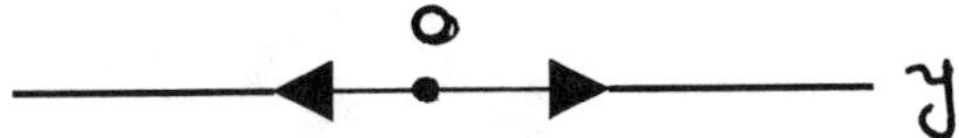

11. The critical points are $P = 0$, 7, and 15. For $P < 0$, $dP/dt > 0$. For $0 < P < 7$, we have $dP/dt < 0$. For $7 < P < 15$, $dP/dt > 0$. Finally, for $P > 15$, $dP/dt < 0$.

The phase portrait is

Since the initial condition falls in the interval $0 < P < 7$, we see that $P(t)$ is decreasing toward the t-axis—that is, $P(t) \to 0$ as $t \to \infty$:

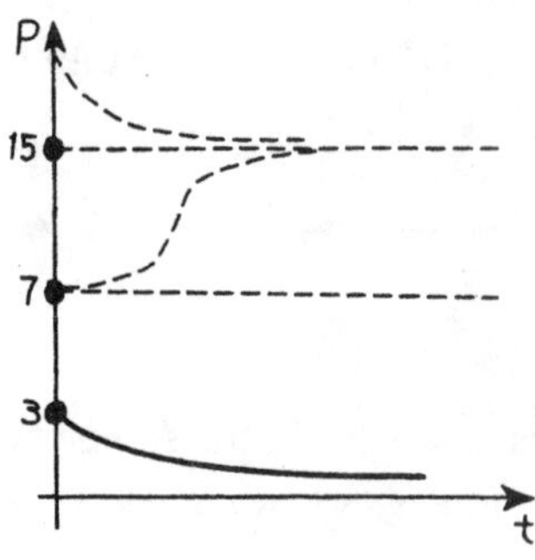

13. (a) We have $\dfrac{dx}{dt} = k(250 - x)(40 - x)$. Setting the derivative equal to zero, we have the critical points $x = 40$ and $x = 250$. If $x < 40$, then $(250 - x) > 0$ and $(40 - x) < 0$, so that $dx/dt > 0$ (remembering that $k > 0$). If $40 < x < 250$, then $(250 - x) > 0$ and $(40 - x) < 0$, so that $dx/dt < 0$. Finally, if $x > 250$, then $(250 - x) < 0$ and $(40 - x) < 0$, so that $dx/dt > 0$. The phase portrait is

(b) If $x(0) = 0$, then $dx/dt > 0$ and x increases toward 40 as $t \to \infty$. (See the slope field for problem 17 in Exercises 2.3.)

15.

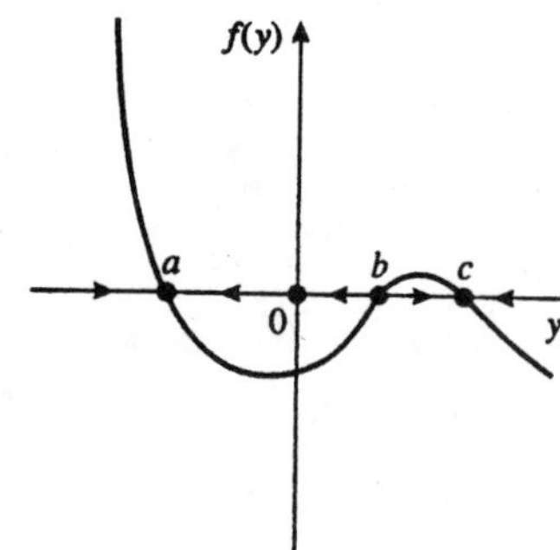

17. (a) $\dfrac{dx}{dt} = ax - bx^3 = x\left(a - bx^2\right) = 0 \Leftrightarrow x = 0, -\sqrt{a/b}$, or $\sqrt{a/b}$. We have four

intervals to examine: $\left(-\infty, -\sqrt{a/b}\right)$, $\left(-\sqrt{a/b}, 0\right)$, $\left(0, \sqrt{a/b}\right)$, and $\left(\sqrt{a/b}, \infty\right)$.

(1) If $x < -\sqrt{a/b}$, then $x < 0$ and $x^2 > a/b$, $bx^2 > a$, $-bx^2 < -a$, $a - bx^2 < 0$, so that $dx/dt > 0$.

(2) If $-\sqrt{a/b} < x < 0$, then $x < 0$ and $x^2 < a/b$, $bx^2 < a$, $-bx^2 > -a$, $a - bx^2 > 0$, so that $dx/dt < 0$.

(3) If $0 < x < \sqrt{a/b}$, then $x > 0$ and $x^2 < a/b$, $bx^2 < a$, $-bx^2 > -a$, $a - bx^2 > 0$, so that $dx/dt > 0$.

(4) When $x > \sqrt{a/b}$, then $x > 0$ and $x^2 > a/b$, $bx^2 > a$, $-bx^2 < -a$, $a - bx^2 < 0$, so that $dx/dt < 0$.

Using the preceding analysis, we can draw the following phase portrait:

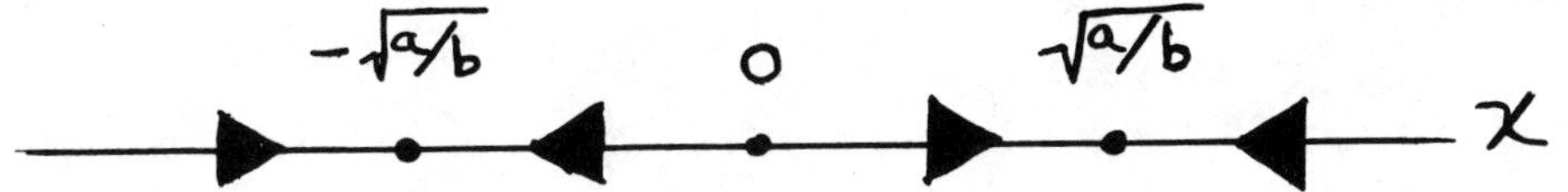

(b) If $x(0)$ is slightly larger than $\sqrt{a/b}$, the phase portrait indicates that $x(t)$ will decrease to $\sqrt{a/b}$ as t increases.

(c) If $x(0) = 0$, then $x(t)$ stays at zero as t increases. Since $dx/dt = 0$ when $x = 0$, we see that there is no change in the value of $x(t)$ at this point.

(d) If $x(0)$ is slightly smaller than $\sqrt{a/b}$, the phase portrait indicates that $x(t)$ will increase to $\sqrt{a/b}$ as t increases.

2.5 Equilibrium Points: Sinks, Sources, and Nodes

< Confirm the answers for Exercises 1 - 12 by looking at slope fields or phase portraits.>

1. $f(y) = y^2(1-y)^2$ and $f'(y) = 2y(2y-1)(y-1)$. The equilibrium points are $y = 0$ and $y = 1$. Now $f'(0) = 0$ and $f'(1) = 0$, so that the Derivative Test fails. However, we see that $f'(y) < 0$ for values of y below 0 and $f'(y) > 0$ for values of y above 0, so that $y = 0$ behaves like both a sink and a source and so is a **node**. Similarly, $f'(y) < 0$ for values of y just below 1 and $f'(y) > 0$ for values above 1, indicating that $y = 1$ is also a **node**.

3. $f(y) = e^y - 1$ and $f'(y) = e^y$. The only equilibrium point is $y = 0$. Since $f'(0) = 1 > 0$, we see that $y = 0$ is a **source**.

5. $f(x) = ax + bx^2$ and $f'(x) = a + 2bx$. The equilibrium points are $x = -a/b$ and 0. Since $f'(-a/b) = -a < 0$, we see that $x = -a/b$ is a **sink**. Since $f'(0) = a > 0$, we conclude that $x = 0$ is a **source**.

7. $f(y) = 10 + 3y - y^2 = (2+y)(5-y)$ and $f'(y) = 3 - 2y$. The equilibrium points are $y = -2$ and 5. Since $f'(-2) = 7 > 0, y = -2$ is a **source**. Since $f'(5) = -7 < 0, y = 5$ is a **sink**.

9. $f(x) = -x^3$ and $f'(x) = -3x^2$. The only equilibrium point is $x = 0$. Since $f'(0) = 0$, we investigate further. Since $f'(x) < 0$ for every nonzero value of x, we conclude that $x = 0$ is a **sink**. The phase portrait confirms this:

11. $f(y) = y\ln(y+2)$ and $f'(y) = y/(y+2) + \ln(y+2)$. The equilibrium points are $y = -1, 0$. Since $f'(-1) = -1 < 0$, we know that $y = -1$ is a **sink**. Since $f'(0) = \ln 2 > 0$, $y = 0$ is a **source**.

13. (a) This problem describes a *compartment* model (see Section 2.2).

$$\frac{dQ}{dt} = D(Q^* - Q) = 0 \Leftrightarrow Q = Q^*.$$

(b) When $Q < Q^*$, $dQ/dt > 0$ and when $Q > Q^*$, $dQ/dt < 0$. Therefore the equilibrium solution is a *sink* and therefore

stable. You could also use the Derivative Test with $f(Q) = D(Q^* - Q)$ and $f'(Q) = -D$, so that $f'(Q^*) = -D < 0$ since D is positive.

15. (a) 

(b) From (a) we see that a is a **sink** and b is a **source**. We can also see this by noticing that $f'(a) < 0$ and $f'(b) > 0$.

17. There is no such equation. Between any two sinks, there must be a source. Suppose there were three sinks, x_1, x_2, and x_3, with $x_1 < x_2 < x_3$. If we consider the phase portrait with just x_1 and x_3 plotted, notice that any point between these two must behave like a *source*:

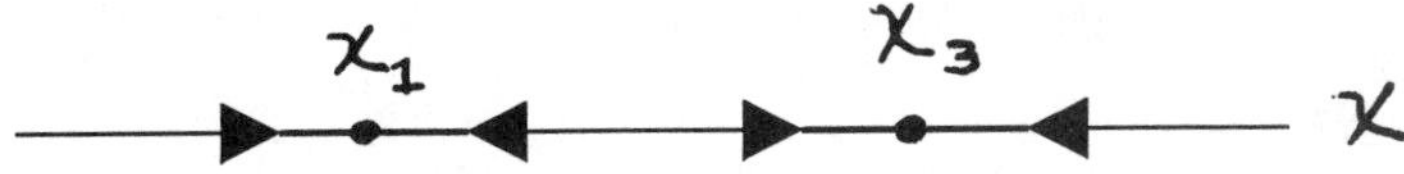

*2.6 Bifurcations

1. (1) 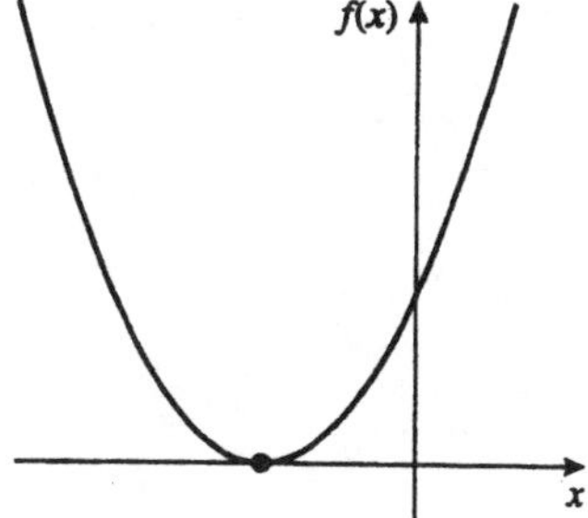

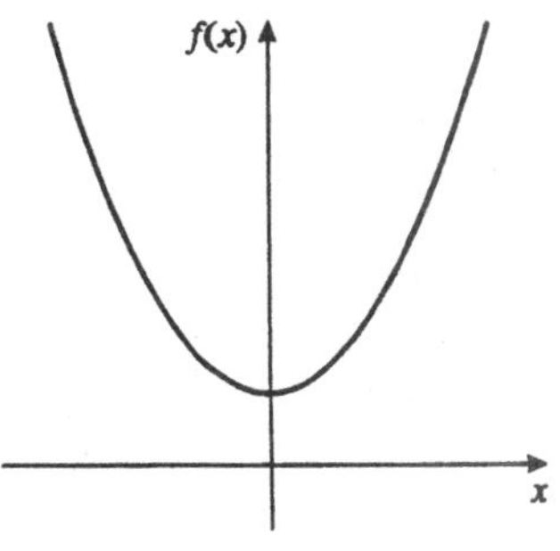

(2) We have $f(x) = 1 + cx + x^2 = 0 \Leftrightarrow x = \dfrac{-c \pm \sqrt{c^2 - 4}}{2}$. Clearly the number and the nature of the equilibrium solution(s) depend on the value of the discriminant $c^2 - 4$. Note that if $c = -2$ or $c = 2$, there is only one equilibrium point, $x = -c/2$. If $c < -2$, then $c^2 - 4 > 0$ and there are two real equilibrium points given by the formula above. When $-2 < c < 2$, we have $c^2 - 4 < 0$, so that there is *no* equilibrium solution. Finally, for $c > 2$, $c^2 - 4 > 0$ and there are two equilibrium solutions again. The bifurcation points are $c = -2$ and $c = 2$.

We note that $f'(x) = c + 2x$ and apply the Derivative Test to determine the nature of the equilibrium solutions: When $c < -2$ or $c > 2$,

$$f'\left(\frac{-c \pm \sqrt{c^2 - 4}}{2}\right) = c + 2\left(\frac{-c \pm \sqrt{c^2 - 4}}{2}\right) = \pm\sqrt{c^2 - 4}$$, so that the

equilibrium solution with the positive square root is a source and the equilibrium solution with the negative square root is a sink. If $c = -2$ or 2, you should check to see that you have a *node*. (The Derivative Test fails to show this.)

(3)

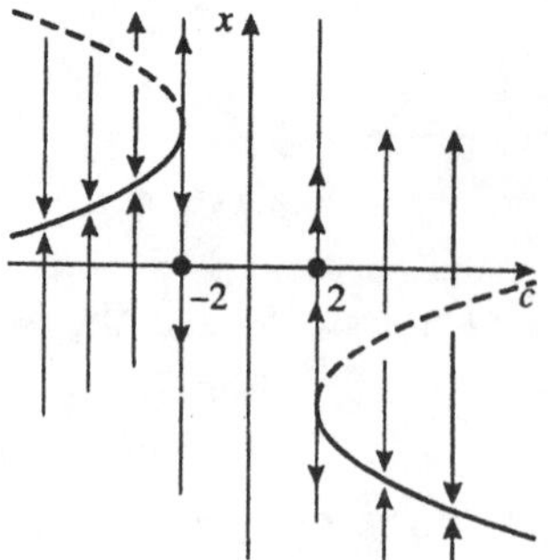

3. (1)

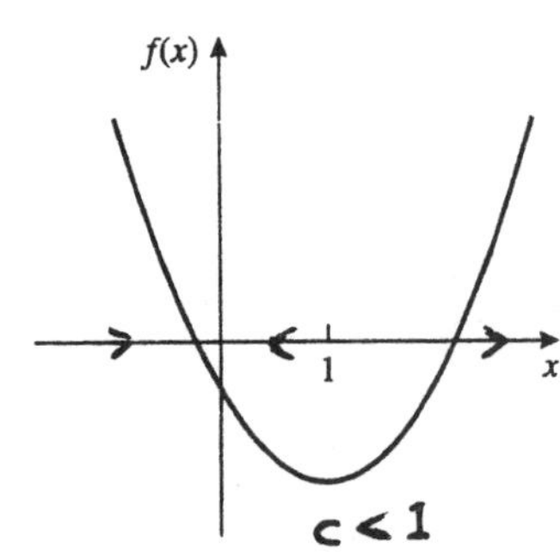

f(x)

1 x

c=1

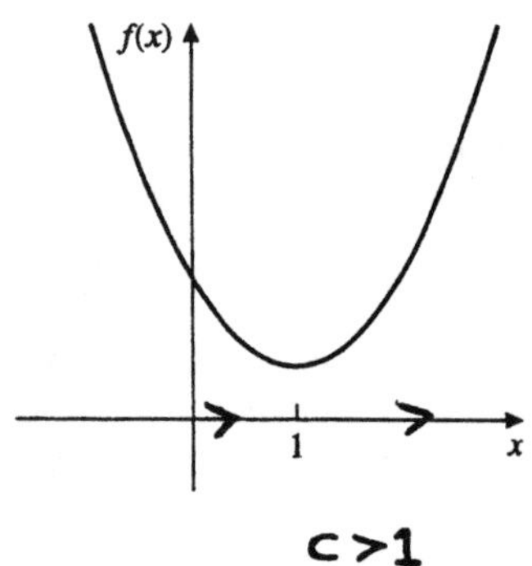

(2) $f(x) = x^2 - 2x + c = 0 \Leftrightarrow x = \dfrac{2 \pm \sqrt{4-4c}}{2} = 1 \pm \sqrt{1-c}$. Therefore, we have two equilibrium points if $c < 1$, one equilibrium point if $c = 1$, and no equilibrium point if $c > 1$. Furthermore, $f'(x) = 2x - 2$, so that $f'\left(1 \pm \sqrt{1-c}\right) = 2\left(1 \pm \sqrt{1-c}\right) - 2 = \pm 2\sqrt{1-c}$. Thus, if $c < 1$, the equilibrium point with the positive square root is a source, while the equilibrium solution with the negative square root is a sink. If $c = 1$, then the equilibrium solution is $x = 1$, a node. Since the qualitative nature of the solutions changes when $c = 1$, we see that $c = 1$ is the only bifurcation point.

(3)

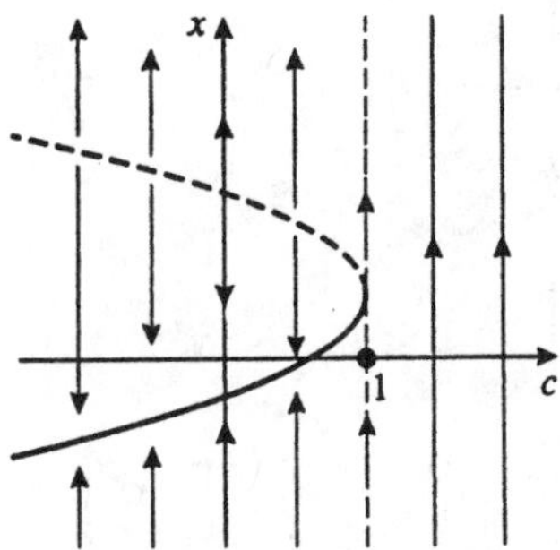

5. $f(x) = x(\alpha - x^2) = 0 \Leftrightarrow x = 0,\ x = -\sqrt{\alpha}$, or $x = \sqrt{\alpha}$. If $\alpha > 0$, there are three equilibrium points. If $\alpha = 0$, there is only one equilibrium point, $x = 0$. If $\alpha < 0$, $x = 0$ is again the only equilibrium point. Therefore $\alpha = 0$ is the only bifurcation point. We have $f'(x) = \alpha - 3x^2$ and we can use the Derivative Test to determine the nature of the equilibrium points. For $\alpha > 0$, $f'(0) = \alpha > 0$, so that $x = 0$ is a source. Also, $f'(-\sqrt{\alpha}) = \alpha - 3\left(-\sqrt{\alpha}\right)^2 = -2\alpha < 0$ and $f'(\sqrt{\alpha}) = \alpha - 3\left(\sqrt{\alpha}\right)^2 = -2\alpha < 0$, indicating that both $-\sqrt{\alpha}$ and $\sqrt{\alpha}$ are sinks. When $\alpha < 0$, $f'(0) = \alpha < 0$, so that $x = 0$ is a sink. When $\alpha = 0$, $f'(0) = 0 - 3(0)^2 = 0$, so that the Derivative Test fails to give us any information. However, the graph of $f(x)$ for $\alpha = 0$, indicates that $x = 0$ is a *sink*:

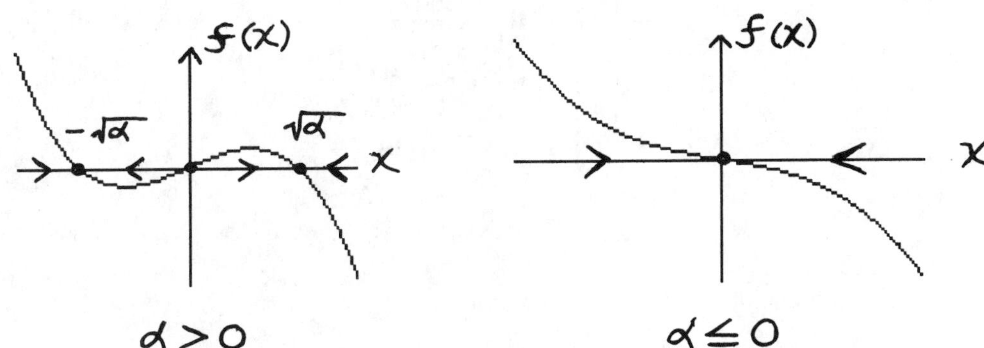

The bifurcation diagram is

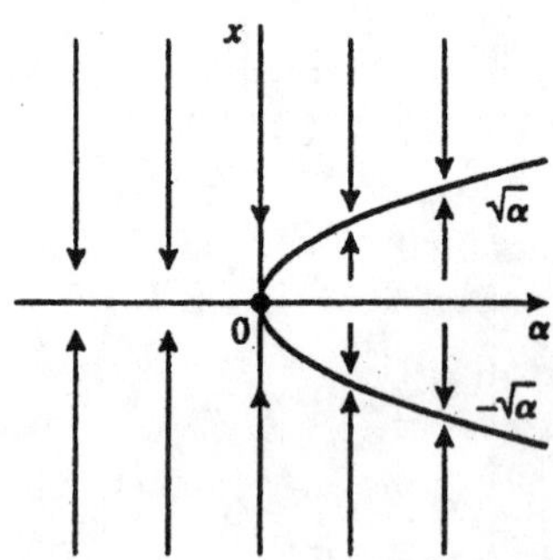

7. (a) $\dot{x} = f(x) = (R - R_c)x - kx^3 = 0 \Leftrightarrow x[(R - R_c) - kx^2] = 0$

$\Leftrightarrow x = 0$ or $x = \pm\sqrt{(R - R_c)/k}$. If $R < R_c$, then $R - R_c < 0$, so that there is only one equilibrium solution, $x = 0$. Since $f'(0) = (R - R_c) - 3k(0)^2$ $= R - R_c < 0$, we see that $x = 0$ is a *sink*.

(b) If $R > R_c$, then $R - R_c > 0$ and we have the three equilibrium solutions $x = 0$, $x = \sqrt{(R - R_c)/k}$, and $x = -\sqrt{(R - R_c)/k}$. Then $f'(0) = (R - R_c) - 3k(0)^2$ $= R - R_c > 0$, showing that $x = 0$ is a source. Also, $f'\left(\pm\sqrt{(R - R_c)/k}\right)$ $= (R - R_c) - 3k\left(\sqrt{(R - R_c)/k}\right)^2 = -2(R - R_c) < 0$, so that $\sqrt{(R - R_c)/k}$ and $-\sqrt{(R - R_c)/k}$ are sinks.

(c)

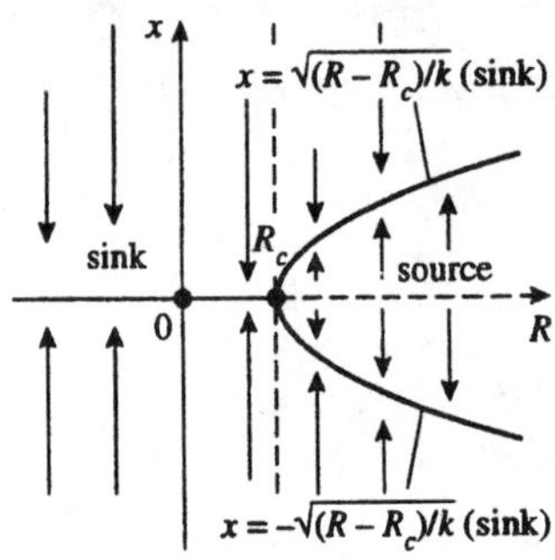

The bifurcation point $R = R_c$ is a *sink*.

2.7 Existence and Uniqueness of Solutions

1. $f(t,x) = \frac{1}{x}$ and $\frac{\partial f}{\partial x} = -\frac{1}{x^2}$ are not continuous where $x = 0$. So, for example, take any rectangle centered at (0, 3) that avoids the t-axis ($x = 0$).

3. $f(t,x) = \frac{x}{t}$ and $\frac{\partial f}{\partial x} = \frac{1}{t}$ are not continuous when $t = 0$—that is, at any point of the x-axis. There is no rectangle R containing the origin that does not also include points of the x-axis, where $t = 0$.

5. $f(t,y) = \frac{t}{1+t+y}$ and $\frac{\partial f}{\partial y} = -\frac{t}{(1+t+y)^2}$ fail to be continuous at those points (t, y) for which $1 + t + y = 0$—that is, at points on the straight line $y = -t - 1$. But the initial condition specifies the point (-2, 1), which lies on this line; and any rectangle that includes (-2, 1) also includes infinitely many points on the line $y = -t - 1$. Clearly, there is no rectangle R satisfying the requirements of the Existence and Uniqueness Theorem.

7. This is a separable equation whose unique solution (using the initial condition) is $y = \tan x$. A look at the graph of this tangent function reveals that the function's domain is $I = (-\pi/2, \pi/2)$, an interval of length π.

9. Separating the variables, we get $x^{-2/3}\,dx = dt$, so that $3x^{1/3} = t + C$, $x^{1/3} = t/3 + K$, and $x(t) = (t/3 + K)^3$. The initial condition implies that $K = x_0$, or $K = \sqrt[3]{x_0}$. Therefore, the solution of the IVP is $x(t) = \left(\frac{t}{3} + \sqrt[3]{x_0}\right)^3$. Note that $f(t,x) = x^{2/3}$ and $\frac{\partial f}{\partial x} = \frac{2}{3\sqrt[3]{x}}$, so that the partial derivative is not continuous at $x = 0$. The initial condition of Example 2.7.2 is $x(0) = 0$, so that we don't expect uniqueness in this case. In the current exercise, both f and $\frac{\partial f}{\partial x}$ are continuous at $(0, x_0)$ if $x_0 < 0$, so that we are guaranteed existence and uniqueness on some t-interval I.

11. (a) We can write the equation as follows:

$$\dot{y} = \begin{cases} \sqrt{y} + k & \text{if } y > 0 \\ \sqrt{-y} + k & \text{if } y < 0 \end{cases}$$

Separating variables in each equation and making the appropriate substitution $u = \sqrt{y} - k$ or $u = \sqrt{-y} - k$, we can integrate to find the implicit solution

$$2\sqrt{y} - 2k\ln\left(\sqrt{y}+k\right) = t+C \text{ if } y>0 \text{ and}$$
$$2\sqrt{|y|} - 2k\ln\left(\sqrt{|y|}+k\right) = t+C \text{ if } y<0.$$

(b) Because $k > 0$, there is a unique solution for *any* initial condition even though $\frac{\partial f}{\partial y}$ does not exist at $y = 0$. As in problem 10 (b), the point is that the Existence and Uniqueness Theorem provides *sufficient* conditions that are not *necessary*. Note that $y \equiv 0$ is not a solution of the equation.

(c) If $k < 0$, the equation has a unique solution for any initial condition. When $k = 0$, we have the solution given in part (a) as well as $y \equiv 0$, so that the IVP with initial condition $y(0) = 0$ has *no* unique solution.

13. Here $f(x,y) = P(x)y^2 + Q(x)y$ and $\frac{\partial f}{\partial y} = 2P(x)y + Q(x)$. Because P and Q are polynomials (continuous everywhere), f and $\frac{\partial f}{\partial y}$ are also continuous everywhere in the x-y plane. The conditions of the Existence and Uniqueness Theorem are satisfied, and so we expect to find an interval $I = (2 - h, 2 + h)$ centered at $x = 2$ such that the IVP has a unique solution on I.

15. $f(P) = kP(b-P)$ and $\frac{\partial f}{\partial P} = kb - 2kP$ are continuous at every point (t, P), so that any IVP involving the logistic equation has a unique solution. If a solution near $P \equiv b$ were to equal the equilibrium solution—that is, if another solution curve intersects the horizontal line $P \equiv b$ at the point (t^*, b)—then we would have *two* solutions of the IVP $\frac{dP}{dt} = kP(b-P)$, $P(t^*) = b$.

17. (a) $f(x,y) = \cos x - x^2y^3$ and $\frac{\partial f}{\partial y} = -3x^2y^2$ are continuous at every point of the x-y plane. Therefore, given a point (x_0, y_0), the equation does have a unique solution passing through (x_0, y_0).

(b) Here's how *Maple* handled the equation:

```
> with(DEtools):
> Eq:=diff(y(x),x)+x^2*(y(x))^2=cos(x);
```

$$Eq := \left(\frac{\partial}{\partial x}\mathrm{y}(x)\right) + x^2\,\mathrm{y}(x)^2 = \cos(x)$$

```
> dsolve(Eq,y(x));
>
```

The last line indicates *Maple*'s response—a blank. This says that *Maple* could not come up with a solution. The moral is that technology doesn't have all the answers. We know from part (a) that every IVP that involves this equation has a unique solution, but we can't seem to find it! However, the following commands (using a numerical method—see Chapter 3—to calculate points on a solution curve) enable us to see some solution curves corresponding to four different initial conditions:

```
> with(DEtools):
> DEplot(diff(y(x),x)+x^2*(y(x))^3=cos(x),y(x),x= - 3..5,{[0,2],[0,-2],[-1,1],[0,0]},
y=-4..4,stepsize=.01,linecolor=black);
```

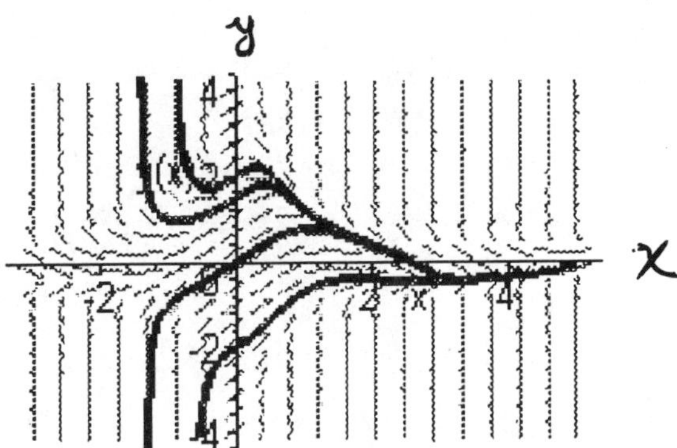

19. If the equation does not have a solution, this may indicate that the reaction cannot take place. However, our confidence in this conclusion must be proportional to our belief that the differential equation provides an accurate description of our experiment. If the reaction *does* take place, we should conclude that the differential equation model is not accurate. On the other hand, if the equation has a solution, this does not guarantee that the reaction does take place, because the model may not be an accurate description of physical reality. For example, in Chapter 4 we'll see a model of a spring-mass system that predicts the mass will never stop bobbing up and down, which certainly is not an accurate description of reality.

Chapter 3

The Numerical Approximation of Solutions

3.1 Euler's Method

1. Here $f(t,y)=t^2-y^2$, $t_0=0$, and $y_0=1$. With $h=0.25$, we have $t_1=0.25$, $t_2=0.50$, $t_3=0.75$, and $t_4=1.00$. Then

$$\begin{aligned}
y_1 &= y_0+0.25f(t_0,y_0)=1+0.25\left(0^2-1^2\right)=0.75\\
y_2 &= y_1+0.25\left(t_1^2-y_1^2\right)=0.75+0.25\left(0.25^2-0.75^2\right)=0.625\\
y_3 &= y_2+0.25\left(t_2^2-y_2^2\right)=0.625+0.25\left(0.50^2-0.625^2\right)=0.58984375\\
y_4 &= y_3+0.25\left(0.75^2-0.58984375^2\right)=0.643489837646\approx 0.643490.
\end{aligned}$$

We can display these approximate values in a table:

t_k	y_k
0	1.000000
0.25	0.750000
0.50	0.625000
0.75	0.589844
1.00	0.643490

Then we can plot the values in the table to give a sketch of the approximate solution curve:

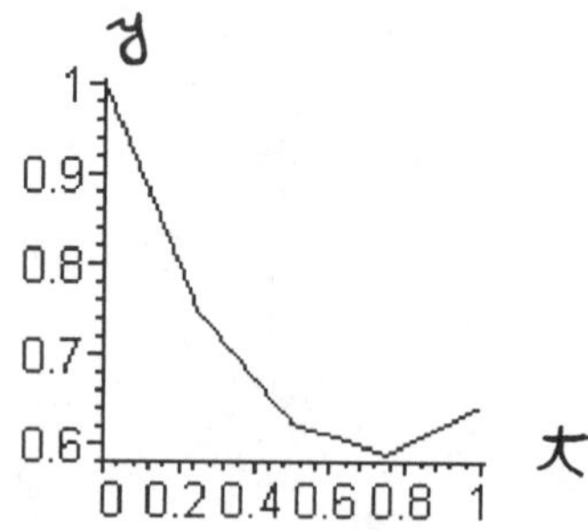

3. In this exercise, we have $f(t,y)=e^{2/y}$, $t_0=1$, and $y_0=2$. With $h=0.5$, we have $t_1=1.5$, $t_2=2.0$, $t_3=2.5$, and $t_4=3.0$. Now

$$y_1=y_0+0.50f(t_0,y_0)=2+0.50\left(e^{2/2}\right)=3.35914091423$$

$$y_2=y_1+0.5\left(e^{2/y_1}\right)=3.35914091423+0.5\left(e^{2/3.35914091423}\right)=4.26601029125$$

$$y_3=y_2+0.5\left(e^{2/y_2}\right)=4.26601029125+0.5\left(e^{2/4.26601029125}\right)=5.06506564414$$

$$y_4=y_3+0.5\left(e^{2/y_3}\right)=5.06506564414+0.5\left(e^{2/5.06506564414}\right)=5.80715503841.$$

The table of approximate values is

t_k	y_k
1.0	2.000000
1.5	3.359141
2.0	4.266010
2.5	5.065065
3.0	5.807155

and the graph of the approximate solution is

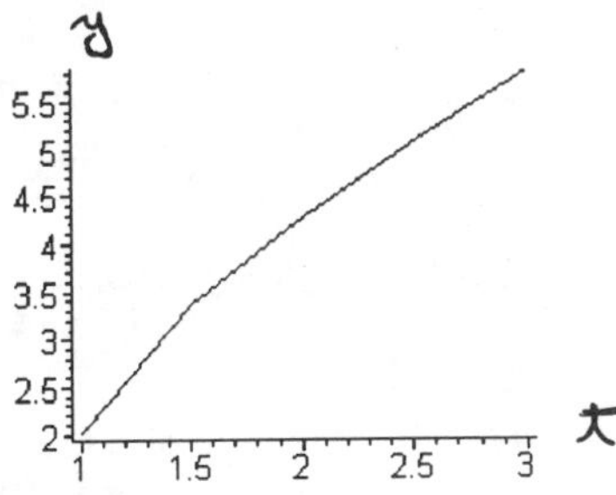

5. We are given $f(t, y)=\cos t$, $t_0=0$, and $y_0=0$. With $h=\pi/10$, we have $t_1=\pi/10$, $t_2=\pi/5$, $t_3=3\pi/10$, $t_4=2\pi/5$, and $t_5=\pi/2$. Then

$$y_1=y_0+\tfrac{\pi}{10}\cos t_0=0+\tfrac{\pi}{10}\cos(0)=\pi/10\approx 0.314159265359$$
$$y_2=y_1+\tfrac{\pi}{10}\cos t_1=0.314159265359+\tfrac{\pi}{10}\cos(\pi/10)\approx 0.612942481833$$
$$y_3=y_2+\tfrac{\pi}{10}\cos t_2=0.612942481833+\tfrac{\pi}{10}\cos(\pi/5)\approx 0.867102666449$$
$$y_4=y_3+\tfrac{\pi}{10}\cos t_3=0.867102666449+\tfrac{\pi}{10}\cos(3\pi/10)\approx 1.0517608495$$
$$y_5=y_4+\tfrac{\pi}{10}\cos t_4=1.0517608495+\tfrac{\pi}{10}\cos(2\pi/5)\approx 1.14884140143$$

A simple integration tells us that $y(t)=\sin t+C$ and $y(0)=0$ implies that $C=0$. Therefore, $y(t)=\sin t$ and $y(\pi/2)=1$. The absolute error of our approximation is

$|1-1.14884140143|=0.14884140143.$

7. We are given $f(x, y)=\dfrac{x}{y}$, $x_0 = 0$, and $y_0 = 1$. With $h = 0.1$, we have $x_k = 0.k$ $(k = 1, 2, \ldots, 10)$.

Then

$$y_1 = y_0 + 0.1\left(\frac{x_0}{y_0}\right) = 1 + 0.1\left(\frac{0}{1}\right) = 1$$

$$y_2 = y_1 + 0.1\left(\frac{x_1}{y_1}\right) = 1 + 0.1\left(\frac{0.1}{1}\right) = 1.01$$

$$y_3 = y_2 + 0.1\left(\frac{x_2}{y_2}\right) = 1.01 + 0.1\left(\frac{0.2}{1.01}\right) = 1.0298019802$$

$$y_4 = y_3 + 0.1\left(\frac{x_3}{y_3}\right) = 1.0298019802 + 0.1\left(\frac{0.3}{1.0298019802}\right) = 1.05893379445$$

$$y_5 = y_4 + 0.1\left(\frac{x_4}{y_4}\right) = 1.05893379445 + 0.1\left(\frac{0.4}{1.05893379445}\right) = 1.09670763849$$

$$y_6 = y_5 + 0.1\left(\frac{x_5}{y_5}\right) = 1.09670763849 + 0.1\left(\frac{0.5}{1.09670763849}\right) = 1.14229864036$$

$$y_7 = y_6 + 0.1\left(\frac{x_6}{y_6}\right) = 1.14229864036 + 0.1\left(\frac{0.6}{1.14229864036}\right) = 1.19482430911$$

$$y_8 = y_7 + 0.1\left(\frac{x_7}{y_7}\right) = 1.19482430911 + 0.1\left(\frac{0.7}{1.19482430911}\right) = 1.25341032838$$

$$y_9 = y_8 + 0.1\left(\frac{x_8}{y_8}\right) = 1.25341032838 + 0.1\left(\frac{0.8}{1.25341032838}\right) = 1.31723619465$$

$$y_{10} = y_9 + 0.1\left(\frac{x_9}{y_9}\right) = 1.31723619465 + 0.1\left(\frac{0.9}{1.31723619465}\right) = 1.38556107091$$

9. (a) We need 100 steps. The same basic *Maple* commands used in Exercise 8 yield $P(1) \approx 1.330624$ million people = 1,330,624 people.

(b) $P(0) = 1.284999...$ million people $\approx$ 1,285,000 people.

11. A CAS using Euler's method with $h = 0.006$ and $V(-0.786) = 150$ yields $V(0) = 166.390541\ldots$ ≈ 166.39 meters per second.

13. A CAS or programmable calculator should yield the following results:

t_k	y_k	**Actual solution**
0	1.000000	1.0000
0.1	1.500000	
0.2	2.190000	
0.3	3.146000	
0.4	4.474400	
0.5	**6.324160**	**8.712004118**
0.6	8.903824	
0.7	12.505354	
0.8	17.537496	
0.9	24.572494	
1.0	**34.411492**	**64.89780316**

The equation is linear and the closed form solution is $y(t) = -\frac{3}{16} + \frac{t}{4} + \frac{19}{16}e^{4t}$, so that $y(0.5) \approx$ 8.712004118. Euler's method has an absolute error of 2.387844118.... Also, $y(1) \approx 64.89780316$, so that the absolute error in this case is 30.4863128816.... The solution curve rises so steeply that the tangent line approximations can't keep up.

15. We are given $f(t,x) = \frac{3t^2}{2x}$, $t_0 = 0$, and $x_0 = 1$. With $h = 0.5$, we have $t_k = 0.5k$, $k = 1, 2, 3, 4$.

Then

$$x_1 = x_0 + 0.5\left(\frac{3t_0^{\ 2}}{2x_0}\right) = 1 + 0.5\left(\frac{3(0^2)}{2(1)}\right) = 1$$

$$x_2 = x_1 + 0.5\left(\frac{3t_1^{\ 2}}{2x_1}\right) = 1 + 0.5\left(\frac{3(0.5)^2}{2(1)}\right) = 1.1875$$

$$x_3 = x_2 + 0.5\left(\frac{3t_2^{\ 2}}{2x_2}\right) = 1.1875 + 0.5\left(\frac{3(1)^2}{2(1.1875)}\right) = 1.81907894737$$

$$x(2) \approx x_4 = x_3 + 0.5\left(\frac{3t_3^{\ 2}}{2x_3}\right) = 1.81907894737 + 0.5\left(\frac{3(1.5)^2}{2(1.81907894737)}\right) = 2.74674621681.$$

With $h = 0.25$, we have $t_k = 0.25k$, $k = 1, 2, 3, 4, 5, 6, 7, 8$. Then

$$x_1 = x_0 + 0.25\left(\frac{3t_0^{\ 2}}{2x_0}\right) = 1 + 0.25\left(\frac{3(0^2)}{2(1)}\right) = 1$$

$$x_2 = x_1 + 0.25\left(\frac{3t_1^{\ 2}}{2x_1}\right) = 1 + 0.25\left(\frac{3(0.25)^2}{2(1)}\right) = 1.0234375$$

$$x_3 = x_2 + 0.25\left(\frac{3t_2^{\ 2}}{2x_2}\right) = 1.0234375 + 0.25\left(\frac{3(0.5)^2}{2(1.0234375)}\right) = 1.11504055344$$

$$x_4 = x_3 + 0.25\left(\frac{3t_3^{\ 2}}{2x_3}\right) = 1.11504055344 + 0.25\left(\frac{3(0.75)^2}{2(1.11504055344)}\right) = 1.30421528735$$

$$x_5 = x_4 + 0.25\left(\frac{3t_4^{\,2}}{2x_4}\right) = 1.30421528735 + 0.25\left(\frac{3(1)^2}{1.30421528735}\right) = 1.59174450406$$

$$x_6 = x_5 + 0.25\left(\frac{3t_5^{\,2}}{2x_5}\right) = 1.59174450406 + 0.25\left(\frac{3(1.25)^2}{1.59174450406}\right) = 1.95985477459$$

$$x_7 = x_6 + 0.25\left(\frac{3t_6^{\,2}}{2x_6}\right) = 1.95985477459 + 0.25\left(\frac{3(1.5)^2}{1.95985477459}\right) = 2.39037136742$$

$$x(2) \approx x_8 = x_7 + 0.25\left(\frac{3t_7^{\,2}}{2x_7}\right) = 2.39037136742 + 0.25\left(\frac{3(1.75)^2}{2.39037136742}\right) = 2.87081449674$$

This is a separable equation with solution $x(t) = \sqrt{t^3 + 1}$. Therefore, $x(2) = 3$. With $h = 0.5$, the absolute error for our approximation of $x(2)$ is $0.25325378319\ldots$; whereas for $h = 0.25$, the absolute error is $0.12918550326\ldots$. The error with $h = 0.25$ is approximately half the error we obtained with $h = 0.50$.

17. The only way a solution curve can coincide with its tangent line segments is if the solution curve is a straight line—that is, if $y(x) = Cx + D$, so that the differential equation is $\frac{dy}{dx} = C$. Suppose that we have the IVP $\frac{dy}{dx} = C$, $y(x_0) = y_0$. Then the solution is given by $\varphi(x) = Cx + (y_0 - Cx_0)$ and it is easy to verify that Euler's method gives us the approximation

$$y_{k+1} = y_k + hC = (y_0 + khC) + hC = y_0 + (k+1)hC.$$

But, with $x_k = x_0 + kh$, we have $\varphi(x_{k+1}) = \varphi(x_0 + (k+1)h) = C(x_0 + (k+1)h) + (y_0 - Cx_0)$ $= y_0 + (k+1)hC = y_{k+1}$. Therefore, the solution's values coincide with the values given by Euler's approximation method.

19. (a) This is a separable equation: $\frac{dy}{dt} = y^{\alpha} \Rightarrow \frac{dy}{y^{\alpha}} = dt \Rightarrow \frac{y^{-\alpha+1}}{1-\alpha} = t + C$. The initial condition implies that $C = 0$, so that $\frac{y^{-\alpha+1}}{1-\alpha} = t$, or $y = [(1-\alpha)t]^{1/(1-\alpha)}$. (Note that $-\alpha + 1 > 0$, so that y remains in the numerator and we never divide by zero.)

(b) We have $f(t, y) = y^{\alpha}$, $t_0 = 0$, and $y_0 = 0$. Choose any step size h. Then

$$\begin{aligned} y_1 &= y_0 + h\,y_0^{\,\alpha} = 0 + h \cdot 0^{\alpha} = 0 \\ y_2 &= y_1 + h\,y_1^{\,\alpha} = 0 + h \cdot 0^{\alpha} = 0 \\ &\vdots \\ y_{k+1} &= y_k + h\,y_k^{\,\alpha} = 0 + h \cdot 0^{\alpha} = 0. \end{aligned}$$

Thus Euler's method converges to the zero function, no matter what value of h is chosen.

(c) If $y_0 = y(0) = 0.01$, then we have, for any h, $y_1 = y_0 + h\,y_0{}^{\alpha} = 0.01 + h(0.01)^{\alpha} \neq 0$. Continuing in this way with Euler's method, we get a nonzero sequence of approximate values. (You may want to use a CAS to do this with $\alpha = 0.5$ and $h = 0.1$, for example.)

21. (a) $\frac{2500}{2501}\cos x + \frac{50}{2501}\sin x - \frac{2500}{2501}e^{-50x}$; $y(0.2) = 0.9836011240\ldots$

(b) $y(0.2) \approx 1.7466146068$. The absolute error is 0.7630134828.

(c) $y(0.2) \approx 1.1761983279$. The absolute error is 0.1925972039.

(d) $y(0.2) \approx 1.8623800769$. The absolute error is 0.8787789529.

(e)

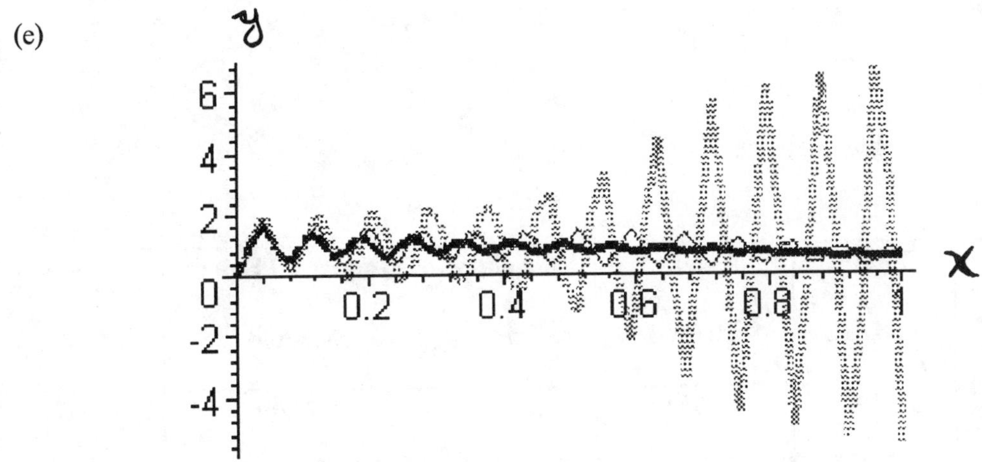

3.3 Improved Euler Method

1. The IVP is $\frac{dx}{dt}=t^2+x,\ x(1)=3$. You are asked to approximate $x(1.5)$. Example 3.1.1 handles the cases $h=0.1$, $h=0.05$, and $h=0.025$ via Euler's method. The case $h=0.1$ is done by hand via the improved Euler method as Example 3.3.2. The following *Maple* commands yield the result for $h=0.05$. The 'stepsize' parameter can be changed to 0.025 to give the remaining result.

```
>SOL1:=dsolve({diff(x(t),t)=t^2+x(t),x(1)=3},x(t),
   type=numeric,stepsize=0.05,method=classical[heunform],startinit=true);
>
                SOL1 := proc(x_classical) ... end proc
> SOL1(1.5);
                [t = 1.5, x(t) = 5.93791426837524749]
```

The table of approximations is

	TRUE VALUE	Euler's Method	Absolute Error	Improved Euler Method	Absolute Error
$h=0.1$	5.93977	5.69513	0.24464	5.93266	0.00711
$h=0.05$	5.93977	5.81260	0.12717	5.93791	0.00186
$h=0.025$	5.93977	5.87490	0.06487	5.93930	0.00047

3. (a) The equation is linear: $\frac{dx}{dt}-x=t$. Using the integration factor $e^{\int -1\,dt}=e^{-t}$, we find

that $\frac{d}{dt}\left[e^{-t}x\right]=t\,e^{-t}$ and $x(t)=-t-1+2e^{t}$.

(b) $x(1)\approx 3.42816$

(c) The following table shows the absolute error at each step of part (b):

t_k	x_k	True Value	Abs. Error
0	1	1	0
0.1	1.11000	1.11034	0.00034
0.2	1.24205	1.24281	0.00076
0.3	1.39847	1.39972	0.00125
0.4	1.58180	1.58365	0.00185
0.5	1.79489	1.79744	0.00255
0.6	2.04086	2.04424	0.00338
0.7	2.32315	2.32751	0.00436
0.8	2.64558	2.65108	0.00550
0.9	3.01236	3.01921	0.00685
1.0	3.42816	3.43656	0.00840

5. Using a CAS and the improved Euler method with $h = 0.006$, we find that

$V(0) \approx 166.27517$ m/s.

3.4 More Sophisticated Numerical Methods: Runge-Kutta and Others

1. The following *Maple* commands will yield an approximation for $y(1) = e$ via the RK4 method:

```
> SOL:=dsolve({diff(y(t),t)=y(t),y(0)=1},y(t),type=numeric,stepsize=0.025,
  method=classical[rk4], startinit=true);
> SOL(1.0);
```

$$SOL := \text{proc}(x\_classical)\ \ldots\ \text{end proc}$$

$$[t = 1.0,\ y(t) = 2.71828181979285333]$$

Using these commands with the required step sizes, we calculate the three approximations:

	TRUE VALUE	Euler's Method	Improved Euler Method	RK4 Method
$h = 0.1$	2.7182818	2.5937425	2.7140808	2.7182797
$h = 0.05$	2.7182818	2.6532977	2.7171911	2.7182817
$h = 0.025$	2.7182818	2.6850638	2.7180039	2.7182818

3. $y(1) = e \approx 2.71828181139414093$. The *Maple* commands are

```
> SOL:=dsolve({diff(y(t),t)=y(t),y(0)=1}, y(t), type=numeric, method=rkf45, startinit=true);
> SOL(1.0);
```

$$SOL := \text{proc}(rkf45\_x) \;...\; \text{end proc}$$

$$[t = 1.0,\ y(t) = 2.71828181139414093]$$

5. (a) The equation is separable: $\frac{dx}{dt} = -tx^2 \Rightarrow \frac{dx}{x^2} = -t\,dt \Rightarrow -x^{-1} = -t^2/2 + K$, or

$$x = \frac{2}{t^2 + C}.$$

(b) If $x(0) = 2$ for the solution given in part (a), then $C = 1$, so that we have $x = \frac{2}{t^2 + 1}$ as the exact solution of the IVP. Then $x(1) = 1$. The rkf45 method yields $x(1) \approx 0.99999999727228860$.

7. (a) The following table provides the required data:

t	$V(t)$
5	100.163
10	104.984
15	105.045
16	105.046
17	105.046
18	105.046
19	105.046
20	105.046

We guess that the terminal velocity is 105.046 ft/sec.

(b) The graph of Ayanna's velocity is

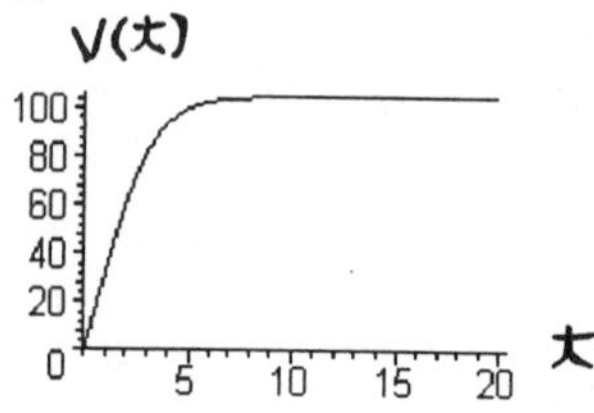

Chapter 4

Second- and Higher-Order Equations

4.1 Homogeneous Second-Order Linear Equations with Constant Coefficients

1. $\lambda^2 - 4\lambda + 4 = 0 \Rightarrow (\lambda - 2)^2 = 0 \Rightarrow \lambda_1 = 2 = \lambda_2 \Rightarrow y = (c_1 + c_2 t)e^{2t}$.

3. $\lambda^2 - 2\lambda + 2 = 0 \Rightarrow \lambda = \dfrac{2 \pm \sqrt{4-8}}{2} = 1 \pm i \Rightarrow x = e^t(c_1 \cos t + c_2 \sin t)$.

5. $\lambda^2 + 2\lambda = 0 \Rightarrow \lambda(\lambda + 2) = 0 \Rightarrow \lambda_1 = 0, \lambda_2 = -2 \Rightarrow x = c_1 + c_2 e^{-2t}$.

7. $\lambda^2 + 4 = 0 \Rightarrow \lambda_1 = 2i, \lambda_2 = -2i \Rightarrow y = c_1 \cos 2t + c_2 \sin 2t$.

9. $\lambda^2 - 4\lambda + 20 = 0 \Rightarrow \lambda = \dfrac{4 \pm \sqrt{16-80}}{2} = 2 \pm 4i \Rightarrow r(t) = e^{2t}(c_1 \cos 4t + c_2 \sin 4t)$.

11. (a)

$$\begin{aligned} L(c_1y_1 + c_2y_2) &= a(c_1y_1 + c_2y_2)'' + b(c_1y_1 + c_2y_2)' + c(c_1y_1 + c_2y_2) \\ &= ac_1y_1'' + ac_2y_2'' + bc_1y_1' + bc_2y_2' + cc_1y_1 + cc_2y_2 \\ &= c_1\left(ay_1'' + by_1' + cy_1\right) + c_2\left(ay_2'' + by_2' + cy_2\right) \\ &= c_1L(y_1) + c_2L(y_2). \end{aligned}$$

(b) $L(c_1 y_1 + c_2 y_2) =$ [by part (a)] $c_1L(y_1) + c_2L(y_2) = c_1 \cdot 0 + c_2 \cdot 0 = 0$.

$\{e^{rt}[ru' + u''] + re^{rt}[ru + u']\} - 2r\{e^{rt}[ru + u']\} + r^2[ue^{rt}] = 0$, or (after simplifying)

13. (a) $Y_1 = \dfrac{y_1 + y_2}{2} = \dfrac{e^{pt}[\cos(qt) + i\sin(qt)] + e^{pt}[\cos(qt) - i\sin(qt)]}{2} = e^{pt}\cos(qt)$, a real-valued function. By the Superposition Principle (p. 127 or Exercise 11(b)], $Y_1 = \frac{1}{2}y_1 + \frac{1}{2}y_2$, as a linear combination of solutions of the homogeneous equation (4.1.1), is a solution of (4.1.1).

(b) $Y_2 = \frac{y_1 - y_2}{2i} = \frac{e^{pt}[\cos(qt) + i\sin(qt)] - e^{pt}[\cos(qt) - i\sin(qt)]}{2i} = e^{pt}\sin(qt)$, a real-valued function. By the Superposition Principle (p. 127 or Exercise 11(b)], $Y_1 = \frac{1}{2i}y_1 - \frac{1}{2i}y_2$, as a linear combination of solutions of the homogeneous equation (4.1.1), is a solution of (4.1.1). The problem statement indicates that complex constants are valid in the Superposition Principle.

(c) Using the results of parts (a) and (b), we have $Y = c_1Y_1 + c_2Y_2 = e^{pt}(c_1\cos(qt) + c_2\sin(qt))$. Because Y_1 and Y_2 are real-valued solutions of the homogeneous equation (4.1.1), the Superposition Principle allows us to conclude that the real-valued linear combination Y is also a solution of (4.1.1). [Re-read the paragraph just before Example 4.1.3 on p. 130.]

15. (a) The equation is $0.1I'' + 6I' + 50 = 0$, or $I'' + 60I' + 500 = 0$, $I(0) = 0, I'(0) = 60$.
The characteristic equation is $\lambda^2 + 60\lambda + 500 = 0$, or $(\lambda + 50)(\lambda + 10) = 0$, so that the eigenvalues are -50 and -10. Thus $I(t) = c_1e^{-50t} + c_2e^{-10t}$. Furthermore, we have $I(0) = 0 = c_1 + c_2$ and $I'(0) = 60 = -50c_1 - 10c_2$, so that $c_1 = -3/2$ and $c_2 = 3/2$. Therefore, $I(t) = -\frac{3}{2}e^{-50t} + \frac{3}{2}e^{-10t}$.

(b) The graph corresponding to the solution found in part (a) is

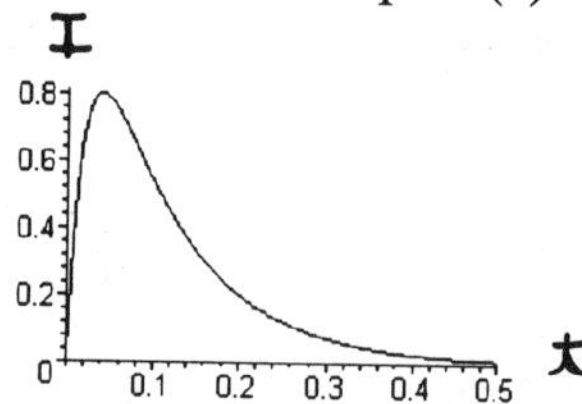

(c) It appears that the maximum value of I is about 0.8. (Your CAS or graphing utility may let you refine this estimate by zooming or tracing.) By setting the derivative of the function found in part (a) equal to 0, we find the critical point $t = (\ln 5)/40 \approx 0.0402359$. It is easy to show that this point yields a maximum value of the function: $I(\ln 5/40) \approx 0.802$.

(d) The maximum value found in part (c) is achieved at $t = (\ln 5)/40 \approx 0.0402359$.

17. $\lambda^2 - 3\lambda + 2 = 0 \Rightarrow (\lambda - 2)(\lambda - 1) = 0 \Rightarrow \lambda_1 = 2, \lambda_2 = 1 \Rightarrow x(t) = c_1e^{2t} + c_2e^t$. The initial conditions give us the simultaneous equations $c_1 + c_2 = 1$ and $2c_1 + c_2 = 0$, with solution $c_1 = -1$, $c_2 = 2$. Therefore, $x(t) = -e^{2t} + 2e^t$.

19. $\lambda^2 - 4\lambda + 20 = 0 \Rightarrow \lambda = \frac{4 \pm \sqrt{16 - 80}}{2} = 2 \pm 4i \Rightarrow y(t) = e^{2t}(c_1\cos 4t + c_2\sin 4t)$. Now $y(\pi/2) = 0$ implies that $c_1 = 0$ and $y'(\pi/2) = 1$ implies that $c_2 = \frac{1}{4}e^{-\pi}$. Therefore, $y(t) = \frac{1}{4}e^{2t-\pi}\sin 4t$.

21. The IVP is $20\ddot{x}+2880x=0,\ x(0)=3,\ \dot{x}(0)=10$. The characteristic equation is $20\lambda^2+2880=0$, with roots $\lambda_{1,2}=\pm 12i$, so that $x(t)=c_1\cos 12t+c_2\sin 12t$. Now $x(0)=3$ implies $c_1=3$ and $\dot{x}(0)=10$ implies $c_2=5/6$. Therefore, $x(t)=3\cos 12t+\frac{5}{6}\sin 12t$.

23. (a) The IVP is $\frac{1}{2}\ddot{x}+2\dot{x}+8x=0$, or $\ddot{x}+4\dot{x}+16x=0,\ x(0)=-0.1,\ \dot{x}(0)=-2$. The characteristic equation is $\lambda^2+4\lambda+16=0$, with roots $\lambda=-2\pm 2\sqrt{3}\,i$, so that $x(t)=e^{-2t}\left(c_1\cos\left(2\sqrt{3}\,t\right)+c_2\sin\left(2\sqrt{3}\,t\right)\right)$. Now $x(0)=-0.1$ implies $c_1=-0.1$ and $\dot{x}(0)=-2$ implies $c_2=-11\sqrt{3}/30$. Therefore, $x(t)=-\frac{1}{30}e^{2t}\left(11\sqrt{3}/30\sin\left(2\sqrt{3}\,t\right)+3\cos\left(2\sqrt{3}\,t\right)\right)$.

(b) The graphs are

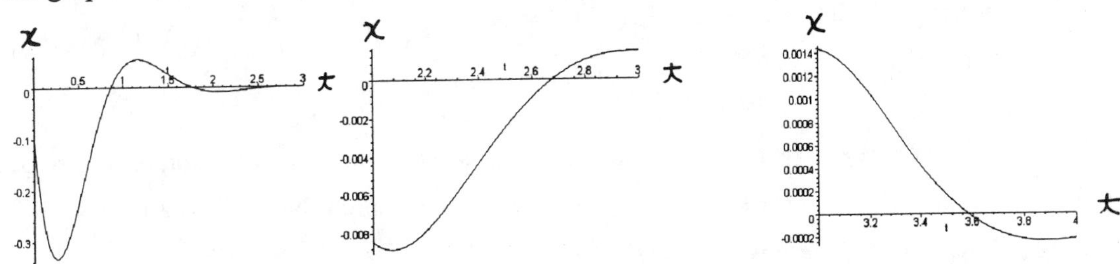

The motion is oscillatory because the damping is light, but the oscillations are decaying.

(c) A closer examination of the first graph in part (b) reveals that the greatest distance of the mass above its equilibrium position is approximately 33 cm.

4.2 Nonhomogeneous Second-Order Linear Equations with Constant Coefficients

1. $y'' - 2y' - 3y = e^{4t}$: The characteristic equation of the associated homogeneous ODE is $\lambda^2 - 2\lambda - 3 = (\lambda - 3)(\lambda + 1) = 0 \Rightarrow \lambda_1 = 3$ and $\lambda_2 = -1 \Rightarrow y_{\text{GH}} = C_1 e^{3t} + C_2 e^{-t}$. Now assume that $y_{\text{PNH}} = y = c_1(t)e^{3t} + c_2(t)e^{-t}$. Then $y' = \left[3c_1 e^{3t} - c_2 e^{-t}\right] + \left[c_1' e^{3t} + c_2' e^{-t}\right.$, and we assume that $c_1' e^{3t} + c_2' e^{-t} = 0$ (*). Next, $y'' = 9c_1 e^{3t} + c_2 e^{-t} + 3c_1' e^{3t} - c_2' e^{-t}$, and we substitute y, y', and y'' in the original nonhomogeneous ODE to conclude that $3c_1' e^{3t} - c_2' e^{-t} = e^{4t}$ (**). Adding (*) and (**), we find that $3c_1' e^{3t} = e^{4t}$, so that $c_1'(t) = \frac{1}{4}e^t$ and $c_1(t) = \frac{1}{4}e^t$. Substituting the expression for c_1' in (*) implies $c_2'(t) = -\frac{1}{4}e^{5t}$ and $c_2(t) = -\frac{1}{20}e^{5t}$. Now $y_{\text{PNH}} = c_1(t)e^{3t} + c_2(t)e^{-t} = \frac{1}{4}e^t e^{3t} - \frac{1}{20}e^{5t}e^{-t} = \frac{1}{4}e^{4t} - \frac{1}{20}e^{4t} = \frac{1}{5}e^{4t}$ and $y_{\text{GNH}} = C_1 e^{3t} + C_2 e^{-t} + \frac{1}{5}e^{4t}$.

3. $x'' - 2x' + 2x = e^t + t\cos t$: The characteristic equation of the associated homogeneous ODE is $\lambda^2 - 2\lambda + 2 = 0 \Rightarrow \lambda = 1 \pm i \Rightarrow x_{\text{GH}} = e^t(C_1 \cos t + C_2 \sin t)$. Now assume that $x_{\text{PNH}} = y = c_1(t)e^t\cos t + c_2(t)e^t\sin t$. Then $x' = \left[(c_2 - c_1)e^t\sin t + (c_1 + c_2)e^t\cos t\right] + \left[c_1' e^t\cos t + c_2' e^t\sin t\right]$, and we assume that $c_1' e^t\cos t + c_2' e^t\sin t = 0$ (*). Next, $x'' = (c_2 - c_1)\left[e^t\cos t + e^t\sin t\right] + (c_2' - c_1')e^t\sin t + (c_1 + c_2)\left[-e^t\sin t + e^t\cos t\right] + (c_1' + c_2')e^t\cos t$. Now substitute x, x', and x'' in the original nonhomogeneous ODE to conclude that $(c_1' + c_2')e^t\cos t + (c_2' - c_1')e^t\sin t$ (**). Using a CAS to solve the equations (*) and (**) simultaneously, we find that $c_1' = -\sin t\left(e^t + t\cos t\right)/e^t$, so $c_1(t) = \cos t - \frac{1}{2}\left\{-\frac{2}{5}t - \frac{4}{25}\right\}e^{-t}\cos(2t) - \frac{1}{2}\left\{-\frac{1}{5}t + \frac{3}{25}\right\}e^{-t}\sin(2t)$. Also, $c_2' = \cos t\left(e^t + t\cos t\right)/e^t$ implies $c_2(t) = \sin t + \frac{1}{2}\left\{-\frac{1}{5}t + \frac{3}{25}\right\}e^{-t}\cos(2t)$ $-\frac{1}{2}\left\{-\frac{2}{5}t - \frac{4}{25}\right\}e^{-t}\sin(2t) - \frac{1}{2}te^{-t} - \frac{1}{2}e^{-t}$. Finally, $x_{\text{PNH}} = \left(-\frac{2}{5}t - \frac{14}{25}\right)\sin t + \left(\frac{1}{5}t + \frac{2}{25}\right)\cos t + e^t$ and $x_{\text{GNH}} = C_1 e^t\cos t + C_2 e^t\sin t + \left(-\frac{2}{5}t - \frac{14}{25}\right)\sin t + \left(\frac{1}{5}t + \frac{2}{25}\right)\cos t + e^t$.

5. $\ddot{x} + \dot{x} = 4\sin t$: The characteristic equation of the associated homogeneous ODE is $\lambda^2 + \lambda = \lambda(\lambda + 1) = 0 \Rightarrow \lambda_1 = 0, \lambda_2 = -1 \Rightarrow x_{\text{GH}} = C_1 + C_2 e^{-t}$. Now assume that $x_{\text{PNH}} = x = c_1(t) + c_2(t)e^{-t}$. Then $\dot{x} = -c_2 e^{-t} + \left(\dot{c}_1 + \dot{c}_2 e^{-t}\right)$, and we assume that $\dot{c}_1 + \dot{c}_2 e^{-t} = 0$ (*). Next, $\ddot{x} = c_2 e^{-t} - c_2' e^{-t}$. Now substitute $\dot{x}$ and $\ddot{x}$ in the original nonhomogeneous ODE to conclude that $-\dot{c}_2 e^{-t} = 4\sin t$ (**). Adding (*) and (**), we find that $\dot{c}_1 = 4\sin t$, so $c_1(t) = -4\cos t$. Then equation (**) yields $\dot{c}_2 = -4e^t\sin t$, which implies $c_2(t) = 2e^t\cos t - 2e^t\sin t$. Finally, $x_{\text{PNH}} = c_1(t) + c_2(t)e^{-t} = -4\cos t + \left[2e^t\cos t - 2e^t\sin t\right]e^{-t} = -2\cos t - 2\sin t$ and $x_{\text{GNH}} = C_1 + C_2 e^{-t} - 2\cos t - 2\sin t$.

7. $y''+4y=2\tan x$: The characteristic equation of the associated homogeneous ODE is $\lambda^2+4=0 \Rightarrow \lambda=\pm 2\,i \Rightarrow y_{\text{GH}}=C_1\cos(2x)+C_2\sin(2x)$. Now assume that $y_{\text{PNH}}=y=c_1(x)\cos(2x)+c_2(x)\sin(2x)$. Then $y'=-2c_1\sin(2x)+2c_2\cos(2x)+c_1'\cos(2x)+c_2'\sin(2x)$, and we assume that $c_1'\cos(2x)+c_2'\sin(2x)=0$ (*).
Next, $y''=-4c_1\cos(2x)-2c_1'\sin(2x)-4c_2\sin(2x)+2c_2'\cos(2x)$. Now substitute y and y'' in the original nonhomogeneous ODE to conclude that $2c_2'\cos(2x)-2c_1'\sin(2x)=2\tan x$ (**).
Using a CAS to solve the equations (*) and (**) simultaneously, we find that $c_1'=-2\sin^2 x$, so $c_1(x)=\cos x\,\sin x-x$. Also, $c_2'=\sin x\,\cos x(2\sin^2 x-1)/(\sin^2 x-1)$ implies $c_2(x)=\sin^2 x+\ln|\cos x|$. Finally, $y_{\text{PNH}}=\cos x\,\sin x\,\cos(2x)-x\cos(2x)+\sin^2 x\,\sin(2x)+\sin(2x)\ln|\cos x|$ and (after using the identity $\sin^2 x=1-\cos^2 x$ and collecting terms) $y_{\text{GNH}}=C_1\cos(2x)+C_2\sin(2x)+\cos x\,\sin x\,\cos(2x)-x\cos(2x)-\cos^2 x\,\sin(2x)+\sin(2x)\ln|\cos x|$.

9. $\ddot{r}+r=1/\sin t$: The characteristic equation of the associated homogeneous ODE is $\lambda^2+1=0 \Rightarrow \lambda=\pm i \Rightarrow r_{\text{GH}}=C_1\cos t+C_2\sin t$. Now assume that $r_{\text{PNH}}=r=c_1(t)\cos t+c_2(t)\sin t$. Then $\dot{r}=-c_1\sin t+c_2\cos t+\dot{c}_1\cos t+\dot{c}_2\sin t$, and we assume that $\dot{c}_1\cos t+\dot{c}_2\sin t=0$ (*). Next, $\ddot{r}=-c_1\cos t-\dot{c}_1\sin t-c_2\sin t+\dot{c}_2\cos t$. Now substitute r, and $\ddot{r}$ in the original nonhomogeneous ODE to conclude that $-\dot{c}_1\sin t+\dot{c}_2\sin t=1/\sin t$ (**). Adding $(\sin t)\times$ (*) and $(-\cos t)\times$ (**), we find that $\dot{c}_1=-1$, so $c_1(t)=-t$. Also, $\dot{c}_2=\cos t/\sin t$ implies $c_2(t)=\ln|\sin t|$. Finally, $r_{\text{PNH}}=-t\cos t+\ln|\sin t|\sin t$ and $r_{\text{GNH}}=C_1\cos t+C_2\sin t-t\cos t+\ln|\sin t|\sin t$.

11. We have the IVP $80\ddot{x}+10000x=2500\cos\left(\frac{\pi v t}{6}\right)$, $x(0)=0$, $\dot{x}(0)=0$. Letting

$x_{\text{PNH}}=A\sin\left(\frac{\pi v t}{6}\right)+B\cos\left(\frac{\pi v t}{6}\right)$, we calculate $\dot{x}_{\text{PNH}}=A\left(\frac{\pi v t}{6}\right)\cos\left(\frac{\pi v t}{6}\right)$ $-B\left(\frac{\pi v t}{6}\right)\sin\left(\frac{\pi v t}{6}\right)$ and $\ddot{x}_{\text{PNH}}=-A\left(\frac{\pi v t}{6}\right)^2\sin\left(\frac{\pi v t}{6}\right)-B\left(\frac{\pi v t}{6}\right)^2\cos\left(\frac{\pi v t}{6}\right)$. Substituting x_{PNH} and $\ddot{x}_{\text{PNH}}$ in the original equation, we wind up with

$$-80A\left(\frac{\pi v t}{6}\right)^2\sin\left(\frac{\pi v t}{6}\right)-80B\left(\frac{\pi v t}{6}\right)^2\cos\left(\frac{\pi v t}{6}\right)+10000A\sin\left(\frac{\pi v t}{6}\right)+10000B\cos\left(\frac{\pi v t}{6}\right)$$

$=2500\cos\left(\frac{\pi v t}{6}\right)$. Setting $t=0$ in this last equation gives us

$-80B\left(\frac{\pi v t}{6}\right)^2+10000B=2500$, or $B\left[-80\left(\frac{\pi v t}{6}\right)^2+10000\right]=2500$, so that we conclude

$B = \dfrac{2500}{10000 - 80\left(\dfrac{\pi v t}{6}\right)^2}$. Setting $t = 3/v$ gives us $-80A\left(\dfrac{\pi v t}{6}\right)^2 + 10000A = 0$, or

$A\left[-80\left(\dfrac{\pi v t}{6}\right)^2 + 10000\right] = 0$, so that $A = 0$. Of course, we assume that

$-80\left(\dfrac{\pi v t}{6}\right)^2 + 10000 \neq 0$. [You can also find A and B by equating coefficients of

$\cos\left(\dfrac{\pi v t}{6}\right)$ and $\sin\left(\dfrac{\pi v t}{6}\right)$ on both sides of the equation.]

13. (a) The IVP is $0.05I'' + 5I' + (1/0.0004)I = -20000\sin(100t)$, $I(0) = 0$, $I'(0) = 4000$. The equation can be rewritten as $I'' + 100I' + 50000I = -400000\sin(100t)$, with characteristic equation $\lambda^2 + 100\lambda + 50000 = 0$ and eigenvalues $-50 \pm 50\sqrt{19}\,i$. Therefore, $I_{GH} = e^{-50t}\left(C_1\cos\left(50\sqrt{19}\,t\right) + C_2\sin\left(50\sqrt{19}\,t\right)\right)$. Rather than use the method of variation of parameters, we can simply guess (*cf.* Examples 4.2.1 and 4.2.2) that a particular solution I_{PNH} of the nonhomogeneous equation has the form $A\cos(100t) + B\sin(100t)$. Substituting this intelligent guess in the nonhomogeneous equation, we find by comparing coefficients of $\cos(100\,t)$ and $\sin(100\,t)$ on both sides of the equation that $A = 40/17$ and $B = -160/17$. Using the initial conditions, we get $C_1 = -40/17$ and $C_2 = 1640\sqrt{19}/323$. Finally, the solution of the IVP is $I(t) = e^{-50t}\left(-\frac{40}{17}\cos\left(50\sqrt{19}\,t\right) + \frac{1640\sqrt{19}}{323}\sin\left(50\sqrt{19}\,t\right)\right)$ $+\ \frac{40}{17}\cos(100t) - \frac{160}{17}\sin(100t)$.

15. Assume that a particular solution of the equation has the form $y_p = A\cos(\omega x) + B\sin(\omega x)$. Then $y_p' = -\omega A\sin(\omega x) + \omega B\cos(\omega x)$ and $y_p'' = -\omega^2 A\cos(\omega x) - \omega^2 B\sin(\omega x)$. Substituting y_p, y_p', and y_p'' in the nonhomogeneous equation yields $\left[-\omega^2 B - 0.2\omega A + B\right]\sin(\omega x)$ $+\left[-\omega^2 A + 0.2\omega B + A\right]\cos(\omega x) = \sin(\omega x)$. Equating coefficients of the sine and cosine terms on each side of this last equation, we get the system of algebraic equations $\left[-\omega^2 B - 0.2\omega A + B\right]\sin(\omega x) = 1$, $\left[-\omega^2 A + 0.2\omega B + A\right]\cos(\omega x) = 0$, with solution $A = -5\omega/\left(25 - 49\omega^2 + 25\omega^4\right)$ and $B = -25\left(1 - \omega^2\right)/\left(25 - 49\omega^2 + 25\omega^4\right)$. Then a particular solution is $y_p = -\dfrac{25\left(1-\omega^2\right)\sin(\omega x)}{25 - 49\omega^2 + 25\omega^4} - \dfrac{5\cos(\omega x)\,\omega}{25 - 49\omega^2 + 25\omega^4}$. Plotting these particular solutions for different values of ω leads to the conclusion that the amplitude of a particular solution is a maximum when $\omega \approx 0.99$. Here are the graphs of y_p for $\omega = 0.5$ (smallest amplitude), 0.99 (largest), and 1.1 (middle):

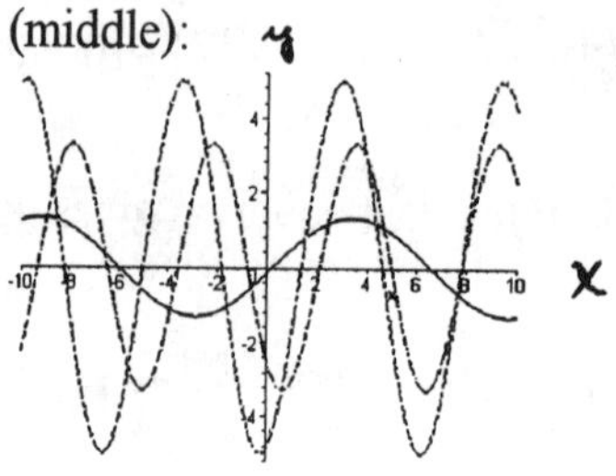

This situation is very close to one in which resonance occurs. However, the oscillations do not build up without limit. In the answers in the back of the text (p. 408), you can see that the general solution of the differential equation has a *transient term* and that any particular solution serves as a *steady-state term* (see Example 4.5.7, for instance).

17. $y'' - 3y' - 4y = 3e^{4x}$: The characteristic equation is $\lambda^2 - 3\lambda - 4 = (\lambda - 4)(\lambda + 1) = 0$, yielding eigenvalues $\lambda_1 = 4$ and $\lambda_2 = -1$. Therefore, $y_{\text{GH}} = C_1 e^{4x} + C_2 e^{-x}$. Ordinarily we would guess at a particular solution of the form Ae^{4x}, but any such solution is already part of y_{GH} and would yield zero when substituted in the original equation. There's a way to handle such a duplication of terms (see p. 246), but we will use the method of variation of parameters in this problem. Therefore, we assume that $y_{\text{PNH}} = y = c_1(x)e^{4x} + c_2(x)e^{-x}$. Then $y' = 4c_1 e^{4x} - c_2 e^{-x}$ $+ c_1' e^{4x} + c_2' e^{-x}$, and we assume that $c_1' e^{4x} + c_2' e^{-x} = 0$ (*). Next we calculate $y'' = 16c_1 e^{4x}$ $+ 4c_1' e^{4x} + c_2 e^{-x} - c_2' e^{-x}$ and substitute y, y', and y'' in the nonhomogeneous equation. This yields $4c_1' e^{4x} - c_2' e^{-x} = 3e^{4x}$ (*). Adding (*) and (* *) gives us $c_1' = \frac{3}{5}$, or $c_1(x) = \frac{3}{5}x$. Substituting in (*) gives us $c_2' = -\frac{3}{5}e^{5x}$, or $c_2(x) = -\frac{3}{25}e^{5x}$, so that $y_{\text{PNH}} = c_1(x)e^{4x} + c_2(x)e^{-x}$ $= \frac{3}{5}xe^{4x} - \frac{3}{25}e^{4x}$ and $y_{\text{GNH}} = C_1 e^{4x} + C_2 e^{-x} + \frac{3}{5}xe^{4x} - \frac{3}{25}e^{4x}$. The initial conditions give us $0 = y(0) = C_1 + C_2 - \frac{3}{25}$ and $0 = 4C_1 - C_2 + \frac{3}{25}$, yielding $C_1 = 0$ and $C_2 = 3/25$. Therefore the solution of the IVP is $y(x) = \frac{3}{5}xe^{4x} - \frac{3}{25}e^{4x} + \frac{3}{25}e^{-x}$.

4.3 Higher-Order Linear Equations with Constant Coefficients

1. $\lambda^4 - 13\lambda^2 + 36 = (\lambda^2 - 4)(\lambda^2 - 9) = (\lambda + 2)(\lambda - 2)(\lambda + 3)(\lambda - 3) = 0 \Rightarrow \lambda_1 = -2,\ \lambda_2 = 2,$
$\lambda_3 = -3,\ \lambda_4 = 3 \Rightarrow y(t) = C_1 e^{-2t} + C_2 e^{2t} + C_3 e^{-3t} + C_4 e^{3t}$.

3. $\lambda^5 + 2\lambda^3 + \lambda = \lambda(\lambda^4 + 2\lambda^2 + 1) = \lambda(\lambda^2 + 1)^2 = 0 \Rightarrow \lambda_1 = 0, \lambda_2 = i = \lambda_3,$
$\lambda_4 = -i = \lambda_5 \Rightarrow y(t) = C_1 + (C_2 \cos t + C_3 \sin t) + t(C_4 \cos t + C_5 \sin t)$.

5. $\lambda^4 - 3\lambda^2 + 2\lambda = \lambda(\lambda^3 - 3\lambda + 2) = \lambda(\lambda - 1)^2(\lambda + 2) = 0 \Rightarrow \lambda_1 = 0, \lambda_2 = 1 = \lambda_3,$
$\lambda_4 = -2 \Rightarrow y(t) = C_1 + (C_2 t + C_3)e^t + C_4 e^{-2t}$.

7. $\lambda^3 - 12\lambda^2 + 22\lambda - 20 = (\lambda - 10)(\lambda^2 - 2\lambda + 2) = 0 \Rightarrow \lambda_1 = 10,$
$\lambda_{2,3} = 1 \pm i \Rightarrow y(t) = C_1 e^{10t} + e^t(C_2 \cos t + C_3 \sin t)$.

9. $y(t) = (C_1 t^2 + C_2 t + C_3)e^t + (C_4 t + C_5)e^{2t} + C_6 e^{3t} + C_7 e^{4t}$.

11. The characteristic equation is $3\lambda^3 + 5\lambda^2 + \lambda - 1 = (\lambda + 1)^2(3\lambda - 1) = 0,$

so that $\lambda_1 = -1 = \lambda_2$ and $\lambda_3 = 1/3$ and the general solution is $y_{\text{GH}} = (C_1 t + C_2)e^{-t} + C_3 e^{t/3}$. The initial conditions yield the system
$\{C_2 + C_3 = 0, C_1 - C_2 + C_3/3 = 1, -2C_1 + C_2 + C_3/9 = -1\}$, with solution
$C_1 = 1/4$, $C_2 = -9/16$, and $C_3 = 9/16$. Therefore, the solution of the IVP is
$y(t) = \left(\frac{t}{4} - \frac{9}{16}\right)e^{-t} + \frac{9}{16}e^{t/3}$.

13. The characteristic equation is $\lambda^5 - \lambda = \lambda(\lambda^2 + 1)(\lambda + 1)(\lambda - 1) = 0$, so that $\lambda_1 = 0$, $\lambda_{2,3} = \pm i$, $\lambda_4 = -1$, and $\lambda_5 = 1$ and the general solution is $y(t) = C_1 + (C_2 \cos t + C_3 \sin t) + C_4 e^{-t} + C_5 e^t$. The initial conditions give us the system
$\left\{\begin{matrix} C_1 + C_2 + C_4 + C_5 = 0, C_3 - C_4 + C_5 = 1, -C_2 + C_4 + C_5 = 0, \\ -C_3 - C_4 + C_5 = 1, C_2 + C_4 + C_5 = 2 \end{matrix}\right\}$,
with solution $C_1 = -2$, $C_2 = 1$, $C_3 = 0 = C_4$, and $C_5 = 1$. Therefore, the solution of the IVP is $y(t) = -2 + \cos t + e^t$.

15. The characteristic equation is $\lambda^2 - 6\lambda + 13 = 0$, so that $\lambda = 3 \pm 2i$, and the general solution of the associated homogeneous equation is $y_{\text{GH}} = e^{3x}(C_1 \cos(2x) + C_2 \sin(2x))$. We guess that $y_{\text{PNH}} = A\cos(2x) + B\sin(2x)$, so that $y'_{\text{PNH}} = -2A\sin(2x) + 2B\cos(2x)$ and $y''_{\text{PNH}} = -4A\cos(2x) - 4B\sin(2x)$. Substituting y_{PNH} and its derivatives in the nonhomogeneous equation, we get $[9A - 12B]\cos(2x) + [12A + 9B]\sin(2x) = 15\cos(2x)$, so that we have the system of equations $9A - 12B = 15$, $12A + 9B = 0$, with solution $A = 3/5$ and $B = -4/5$. Therefore, $y_{\text{GNH}} = e^{3x}(C_1 \cos(2x) + C_2 \sin(2x)) + \frac{3}{5}\cos(2x) - \frac{4}{5}\sin(2x)$.

17. The characteristic equation of the homogeneous equation is $\lambda^3 + 6\lambda^2 + 11\lambda + 6 = (\lambda + 3)(\lambda + 2)(\lambda + 1) = 0$, so that $\lambda_1 = -3$, $\lambda_2 = -2$, and $\lambda_3 = -1$. Therefore, the general solution of the homogeneous equation is $y_{\text{GH}} = C_1 e^{-3x} + C_2 e^{-2x} + C_3 e^{-x}$. Then $y_{\text{PNH}} = c_1(x)e^{-3x} + c_2(x)e^{-2x} + c_3(x)e^{-x}$. After differentiating, substituting, and solving for c'_i ($i = 1, 2, 3$), we find that $c_1(x) = xe^{3x}$, $c_2(x) = -3xe^{2x}$, and $c_3(x) = 3xe^x - \frac{13}{2}e^x$. Then $y_{\text{PNH}} = x - 3$ and $y_{\text{GNH}} = x - 3 + C_1 e^{-3x} + C_2 e^{-2x} + C_3 e^{-x}$.

19. The characteristic equation of the homogeneous equation is $\lambda^3 + 5\lambda^2 - 6\lambda = \lambda(\lambda + 6)(\lambda - 1) = 0$, so that $\lambda_1 = 0$, $\lambda_2 = -6$, and $\lambda_3 = 1$. Therefore, the general solution of the homogeneous equation is $y_{\text{GH}} = C_1 + C_2 e^{-6x} + C_3 e^x$. Then $y_{\text{PNH}} = c_1(x) + c_2(x)e^{-6x} + c_3(x)e^x$. After differentiating, substituting, and solving for c'_i ($i = 1, 2, 3$), we find that $c_1(x) = -\frac{1}{14}e^x$, $c_2(x) = \frac{1}{98}e^{7x}$, and $c_3(x) = \frac{3}{7}x$. Then $y_{\text{PNH}} = -\frac{3}{49}e^x + \frac{3}{7}xe^x$ and $y_{\text{GNH}} = \frac{3}{7}xe^x + C_1 + C_2 e^{-6x} + C_3 e^x$. (Note that one term of the particular solution has been absorbed into the general solution.) Using the initial conditions, we find that $C_1 = 1$, $C_2 = 0$, and

$C_3 = 0$, so that $y_{\text{GNH}} = \frac{3}{7} x e^x + 1$.

4.4 Higher-Order Equations and Their Equivalent Systems

1. Let $x_1 = x$ and $x_2 = \frac{dx}{dt} = \frac{dx_1}{dt}$. Then $\frac{dx_2}{dt} = \frac{d^2x}{dt^2} = 1 + x = 1 + x_1$ and we have the system $\left\{\frac{dx_1}{dt} = x_2, \frac{dx_2}{dt} = 1 + x_1\right\}$.

3. $x_1 = x$ and $x_2 = x' \Rightarrow x_1' = x_2$ and $x_2' = x'' = 1 - 3x' - 2x = 1 - 3x_2 - 2x_1$. Also, $x(0) = 1$ implies that $x_1(0) = 1$ and $x'(0) = 0$ implies that $x_2(0) = 0$. The system is $\{x_1' = x_2,\ x_2' = 1 - 3x_2 - 2x_1;\ x_1(0) = 1,\ x_2(0) = 0\}$.

5. $w_1 = w, w_2 = w', w_3 = w''$, and $w_4 = w''' \Rightarrow w_1' = w_2,\ w_2' = w'' = w_3, w_3' = w''' = w_4$, and $w_4' = w^{(4)} = 6\sin(4t) + 2w''' - 5w'' - 3w' + 8w = 6\sin(4t) + 2w_4 - 5w_3 - 3w_2 + 8w_1$. The nonautonomous system is $\{w_1' = w_2, w_2' = w_3, w_3' = w_4, w_4' = 6\sin(4t) + 2w_4 - 5w_3 - 3w_2 + 8w_1\}$. To get an *autonomous* system, replace t by w_5 and add the equation $w_5' = 1$.

7. $y_1 = y$ and $y_2 = y' \Rightarrow y_1' = y' = y_2$ and $y_2' = y'' = (5\ln x + 3xy' - 4y)/x^2$ $= (5\ln x + 3xy_2 - 4y_1)/x^2$. The nonautonomous system is $\{y_1' = y_2, y_2' = (5\ln x + 3xy_2 - 4y_1)/x^2\}$. Replacing x by y_3 and adding the equation $y_3' = 1$ yields an autonomous system.

9. Let $x_1 = x$, $x_2 = \frac{dx}{dt}$, $y_1 = y$, and $y_2 = \frac{dy}{dt}$. Then $\frac{dx_1}{dt} = x_2$, $\frac{dx_2}{dt} = \frac{d^2x}{dt^2} = -x = -x_1$, $\frac{dy_1}{dt} = y_2$, and $\frac{dy_2}{dt} = \frac{d^2y}{dt^2} = y = y_1$. The system is $\left\{\frac{dx_1}{dt} = x_2, \frac{dx_2}{dt} = -x_1, \frac{dy_1}{dt} = y_2, \frac{dy_2}{dt} = y_1\right\}$.

11. Let $y_1 = y$ and $y_2 = \frac{dy}{dt}$. Then $\frac{dy_1}{dt} = \frac{dy}{dt} = y_2$ and $\frac{dy_2}{dt} = \frac{d^2y}{dt^2} = -\left(\frac{g}{s_0}\right)y = -\left(\frac{g}{s_0}\right)y_1$. The system is $\left\{\frac{dy_1}{dt} = y_2, \frac{dy_2}{dt} = -\left(\frac{g}{s_0}\right)y_1\right\}$.

13. If $y_1 = y$, $y_2 = y'$, and $y_3 = y''$, then $y_1' = y_2, y_2' = y_3$, and $y_3' = y''' = \cos y - y' = \cos y_1 - y_2$. The system is $\{y_1' = y_2, y_2' = y_3, y_3' = \cos y_1 - y_2\}$.

15. The nonautonomous system for Example 4.4.5 can be converted to an autonomous system by letting $u_3 = x$, so that $u_3(0) = 0$ and $u_3'(0) = 1$. Then the system becomes $\{u_1' = u_2, u_2' = u_3 u_2 + u_3^2 u_1,\ u_3' = 1;\ u_1(0) = 1, u_2(0) = 2, u_3(0) = 0\}$.

17. Let $x_1 = x$, $x_2 = \frac{dx}{dt}$, $y_1 = y$, and $y_2 = \frac{dy}{dt}$. Then $\frac{dx_1}{dt} = x_2$, $\frac{dx_2}{dt} = \frac{d^2x}{dt^2} = \frac{1}{2}\left\{x - \left(\frac{dy}{dt}\right)^2\right\}$ $= \frac{1}{2}\{x_1 - y_2^2\}$, $\frac{dy_1}{dt} = y_2$, and $\frac{dy_2}{dt} = \frac{d^2y}{dt^2} = \frac{(4t+y)}{x} = \frac{4t+y_1}{x_1}$.

19. Let $w = \ln t$. Then, using the Chain Rule (Appendix A.2), we see that $\frac{dx}{dt} = \frac{dx}{dw}\cdot\frac{dw}{dt} = \frac{1}{t}\cdot\frac{dx}{dw}$, so that $t\frac{dx}{dt} = -3x + 4y$ becomes $t\left(\frac{1}{t}\frac{dx}{dw}\right) = -3x + 4y$, or $\frac{dx}{dw} = -3x + 4y$. Similarly, $\frac{dy}{dt} = \frac{dy}{dw}\cdot\frac{dw}{dt} = \frac{1}{t}\frac{dy}{dw}$ and we get $t\frac{dy}{dt} = \frac{dy}{dw} = -2x + 3y$.

4.5 Qualitative Analysis of Autonomous Systems

1. (a) $x_1 = x, x_2 = x' \Rightarrow x_1' = x_2, x_2' = x'' = -x' = -x_2; x(0) = 1 \Rightarrow x_1(0) = 1$, $x'(0) = 2 \Rightarrow x_2(0) = 2$. The system is $\{x_1' = x_2, x_2' = -x_2; x_1(0) = 1, x_2(0) = 2\}$.

   (b) The graph of the solution in the phase plane is

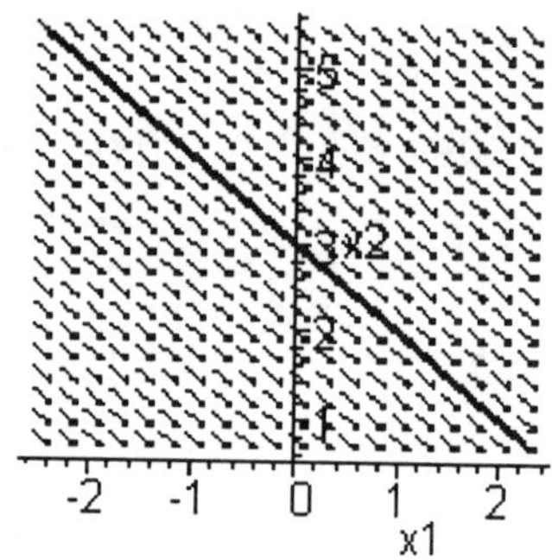

   (c) The graphs of the solutions x_1 and x_2 relative to the t-axis are

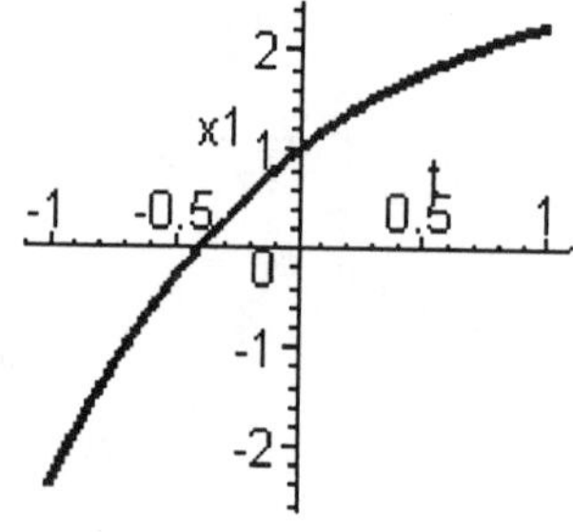

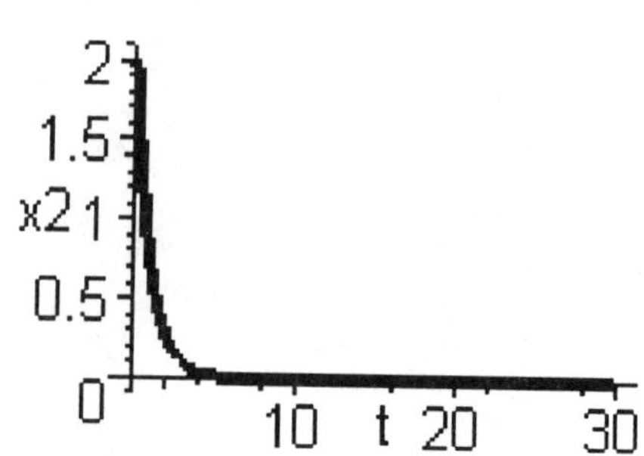

3. (a) $y_1 = y, y_2 = \dot{y} \Rightarrow \dot{y}_1 = y_2, \dot{y}_2 = \ddot{y} = -y = -y_1\,; y(0) = 2 \Rightarrow y_1(0) = 2,$
$\dot{y}(0) = 0 \Rightarrow y_2(0) = 0$. The system is $\{\dot{y}_1 = y_2, \dot{y}_2 = -y_1\,; y_1(0) = 2, y_2(0) = 0\}$.

(b) The graph of the solution in the phase plane is

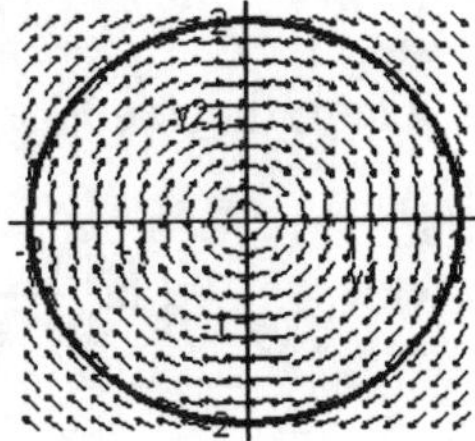

(c) The graphs of y_1 and y_2 relative to the t-axis are

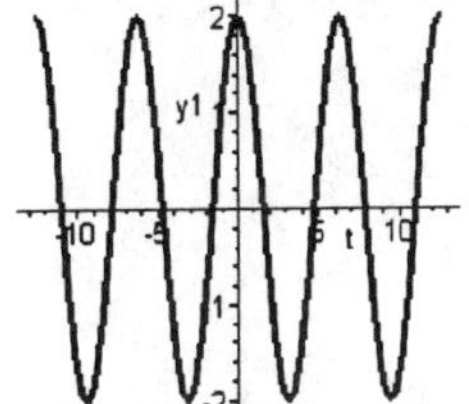

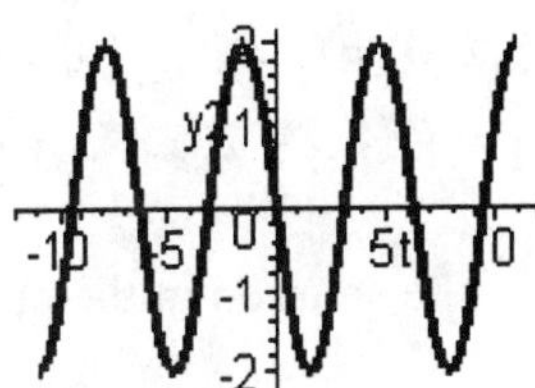

5. (a) $x_1 = x, x_2 = \dot{x} \Rightarrow \dot{x}_1 = x_2, \dot{x}_2 = \ddot{x} = \dot{x} = x_2\,; x(0) = 1 \Rightarrow x_1(0) = 1, \dot{x}(0) = 1 \Rightarrow x_2(0) = 1$. The system is $\{\dot{x}_1 = x_2, \dot{x}_2 = x_2\,; x_1(0) = 1 = x_2(0)\}$.

(b) The graph of the solution in the phase plane is

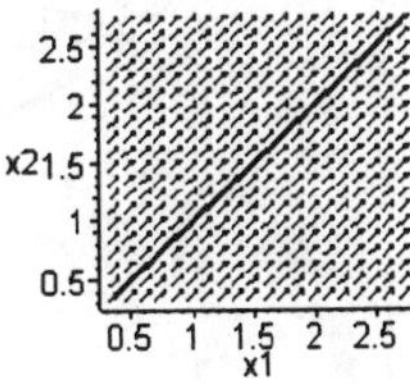

(c) The graphs of x_1 and x_2 against t are

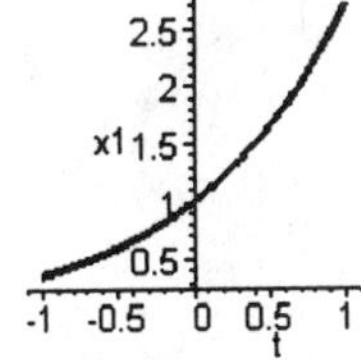

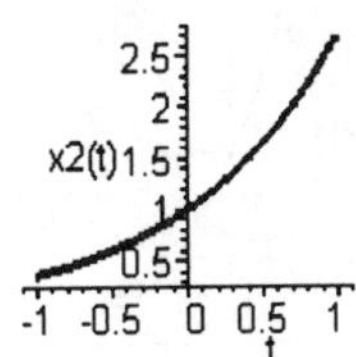

7. $\frac{dy}{dx}=\frac{-x+y}{x+y}=\frac{x\left(-1+\frac{y}{x}\right)}{x\left(1+\frac{y}{x}\right)}=\frac{-1+\frac{y}{x}}{1+\frac{y}{x}}$. Let $z=y/x$, so that $y=xz$ and $\frac{dy}{dx}=x\frac{dz}{dx}+z$.

Then the equation becomes $x\frac{dz}{dx}+z=\frac{-1+z}{1+z}$, $x\frac{dz}{dx}=-\frac{1+z^2}{1+z}$, $\frac{1+z}{1+z^2}\,dz=-\frac{1}{x}dx$,

$\left(\frac{1}{1+z^2}+\frac{z}{1+z^2}\right)dz=-\frac{1}{x}dx$, so that by integrating we have

$\arctan z+\frac{1}{2}\ln\left|1+z^2\right|=-\ln|x|+C$, or $\arctan\left(\frac{y}{x}\right)+\frac{1}{2}\ln\left(\frac{x^2+y^2}{x^2}\right)+\ln|x|-C=0$. This last

equation can be further simplified to $2\arctan\left(\frac{y}{x}\right)+\ln\left(x^2+y^2\right)-C=0$.

9. (a) $x_1=x,\ x_2=\dot{x}\ \Rightarrow\ \dot{x}_1=x_2,\ \dot{x}_2=\ddot{x}=-64x-20\dot{x}=-64x_1-20x_2$; $x(0)=1/3\ \Rightarrow$
$x_1(0)=1/3, \dot{x}(0)=0\ \Rightarrow\ x_2(0)=0$. The system is
$\{\dot{x}_1=x_2, \dot{x}_2=-64x_1-20x_2;\ x_1(0)=1/3, x_2(0)=0\}$.

(b) The graph of the solution in the phase plane is

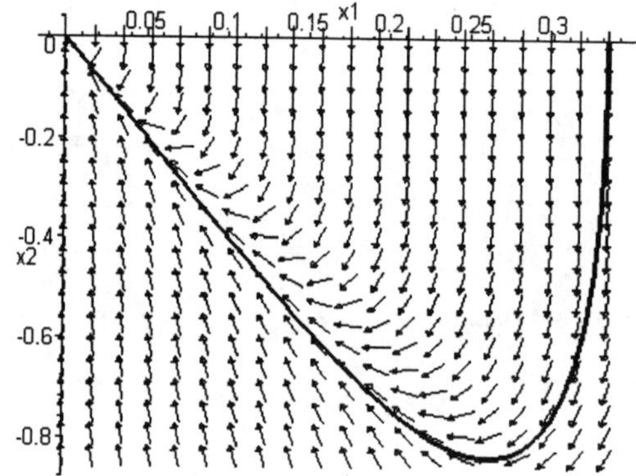

(c) The graph of $x(t)$ $(=x_1(t))$ against t is

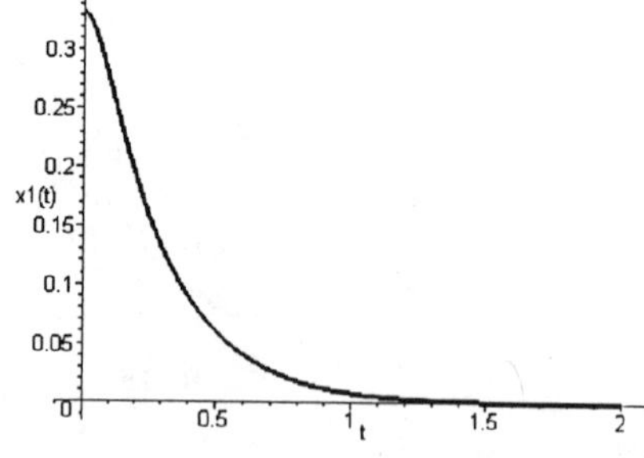

(d) The mass approaches its equilibrium position, but doesn't quite reach it because of the large damping force. In particular, the mass doesn't overshoot its equilibrium position.

11. (a) The system is $\{\dot{x}_1 = x_2, \dot{x}_2 = 16\cos 8t - 64x_1;\ x_1(0) = 0, x_2(0) = 0\}$.

(b) The graph of the solution in the phase plane is

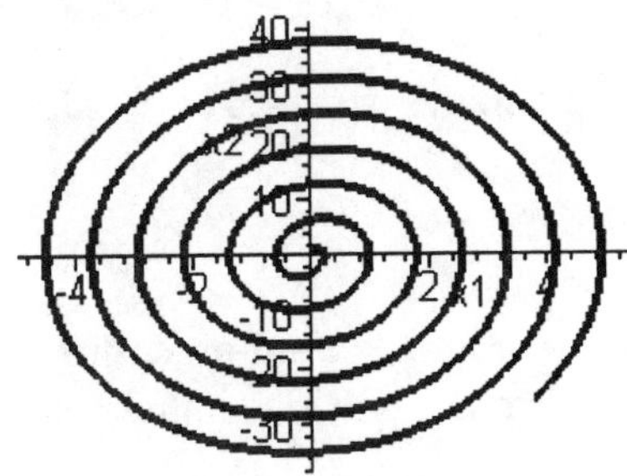

(c) The graph of $x(t)$ $(= x_1(t))$ against t is

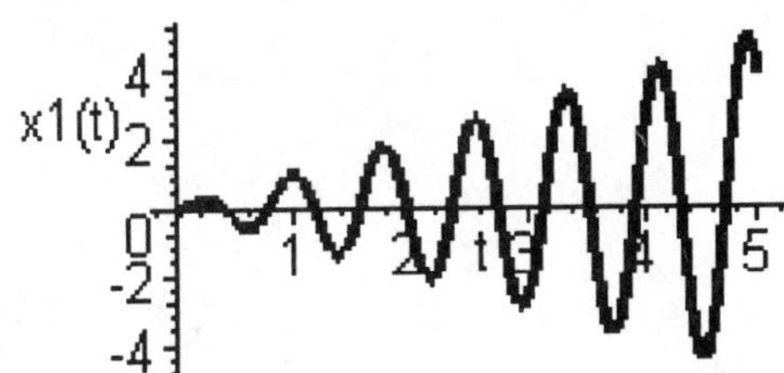

(d) The half lines $x = t$ and $x = -t$ are asymptotes for the graph in (c) for $t \geq 0$:

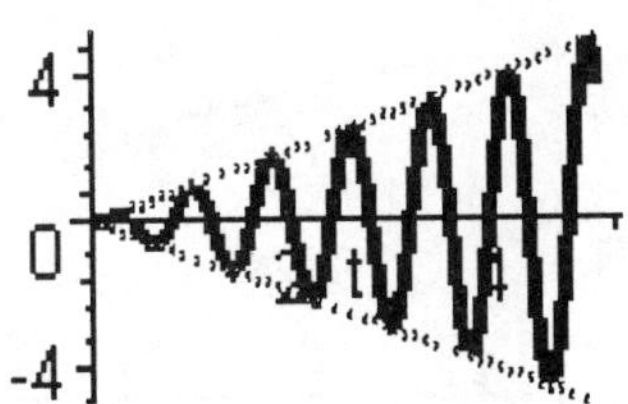

13. (a) $Q'' = -6Q' + 3R' = -6Q' + 3[-Q - 2R] = -6Q' - 3Q - 6R = -6Q' - 3Q - 6\left[\dfrac{Q' + 6Q}{3}\right]$

$= -8Q' - 15Q$, or $Q'' + 8Q' + 15Q = 0$. This equation *can* represent a spring-mass system with spring constant 15 and damping constant 8. (We could also have derived a single second-order equation in R.)

(b) $\ddot{x} = 3\dot{x} - \dot{y} = 3\dot{x} - [x + 3y] = 3\dot{x} - x - 3y = 3\dot{x} - x - 3[3x - \dot{x}] = 6\dot{x} - 10x$,

or $\ddot{x} - 6\dot{x} + 10x = 0$. This second-order equation *cannot* represent a spring-mass system because the equation implies that any damping force works in the same direction as the mass's motion. (We could also have derived a single second-order equation in y.)

15. The system is $\{x_1' = x_2,\ x_2' = x_1 - x_1^3 - x_2\}$. Then we have $x_2 = 0$ and $x_1 - x_1^3 - x_2 = 0$. Substituting $x_2 = 0$ into the second equation gives us $x_1 - x_1^3 = x_1(1 + x_1)(1 - x_1) = 0$, so that $x_1 = 0$, -1, or 1. Therefore, the equilibrium solutions are (0, 0), (-1, 0), and (1, 0).

4.6 Existence and Uniqueness

1. $x(t)=\frac{1}{2}\left(e^{t}+e^{-t}\right)$, $x'=\frac{1}{2}\left(e^{t}-e^{-t}\right)$, $x''=\frac{1}{2}\left(e^{t}+e^{-t}\right)$. Then $x''-x=$ $\frac{1}{2}\left(e^{t}+e^{-t}\right)-\frac{1}{2}\left(e^{t}+e^{-t}\right)=0$. Furthermore, $x(0)=\frac{1}{2}\left(e^{0}+e^{-0}\right)=1$ and $x'(0)=\frac{1}{2}\left(e^{0}-e^{-0}\right)=0$.

3. The Existence and Uniqueness Theorem for a second-order differential equation requires *two* initial conditions. The equation in this problem has only one initial condition, so one shouldn't expect the theorem to apply.

5. (a) $\dfrac{dx_1}{dt}=3e^{-t}\cos(3t)-e^{-t}\sin(3t)=3y_1-x_1=-x_1+3y_1$ and

$\dfrac{dy_1}{dt}=-3e^{-t}\sin(3t)-e^{-t}\cos(3t)=-3x_1-y_1$. Similarly,

$\dfrac{dx_2}{dt}=3e^{-(t-1)}\cos(3(t-1))-e^{-(t-1)}\sin(3(t-1))=3y_2-x_2=-x_2+3y_2$ and

$\dfrac{dy_2}{dt}=-3e^{-(t-1)}\sin(3(t-1))-e^{-(t-1)}\cos(3(t-1))=-3x_2-y_2$.

(b) The graph of $(x_1(t), y_1(t))$ in the x-y phase plane is

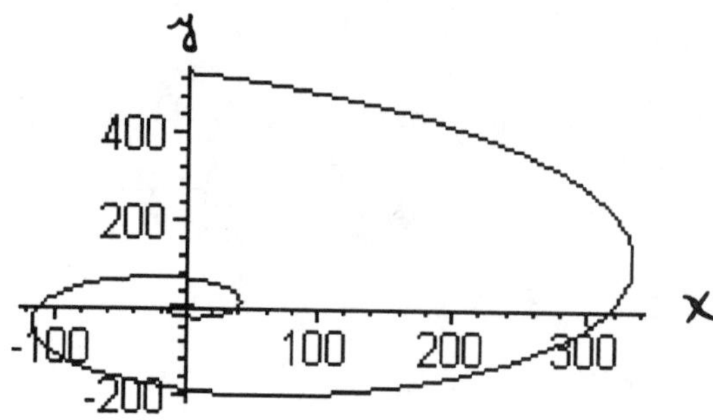

The graph of $(x_2(t), y_2(t))$ in the x-y phase plane is

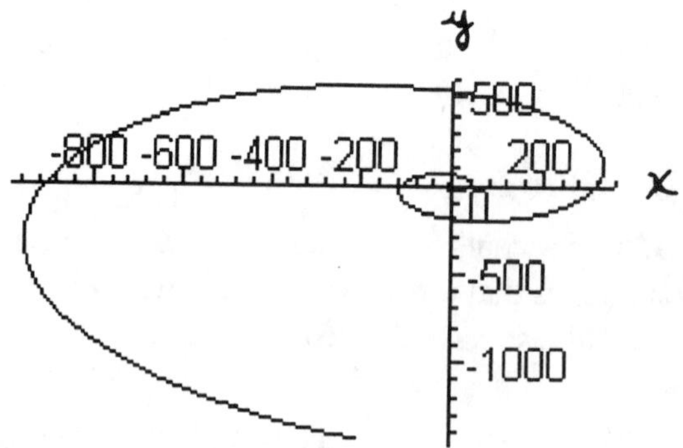

(c) Both graphs represent the same spiral. This does not contradict the uniqueness part of our theorem because they are never in the same place at the same time—that is, for any particular value of t, say t^*, we have $\left(x_1(t^*), y_1(t^*)\right) \neq \left(x_2(t^*), y_2(t^*)\right)$.

7. Using the initial points $(x(0), y(0)) = (1, 2), (-1, 2), (-1, -2)$, and $(1, -2)$, for example, we get the trajectories

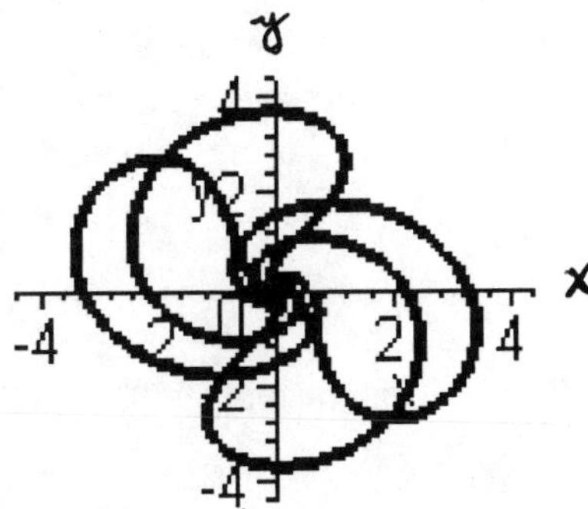

This situation does not contradict the existence and uniqueness theorem because two distinct trajectories are never in the same place *at the same time*. In t-x-y space, the graphs of different solutions will have no points in common. because the first coordinate t will be different at each distinct point of the graph.

4.7 Numerical Solutions

1. (a) $x_{k+1} = x_k + \frac{h}{2}\{f(t_k, x_k, y_k) + f(t_{k+1}, x_k + hf(t_k, x_k, y_k), y_k + hg(t_k, x_k, y_k))\}$

$y_{k+1} = y_k + \frac{h}{2}\{g(t_k, x_k, y_k) + g(t_{k+1}, x_k + hf(t_k, x_k, y_k), y_k + hg(t_k, x_k, y_k))$

(b) $x(0.5) \approx 1.1273$, $y(0.5) \approx 0.5202$.

(c) For $x(0.5)$, the absolute error is approximately 0.0003; while for $y(0.5)$, the absolute error is approximately 0.0009.

(c) Using a fourth-order Runge-Kutta method with $h = 0.2$, we get $x(0.2) \approx 1.320067$, with absolute error 0.00035777, and $y(0.2) \approx -0.250667$, with absolute error 0.000180012.

3. (a) The system is $\{u_1' = u_2,\ u_2' = 2x + 2u_1 - u_2;\ u_1(0) = 1,\ u_2(0) = 1\}$.

(b) Using Euler's method with $h = 0.1$, we find that $u_1(0.5) \approx 1.8774$ and $u_2(0.5) \approx 4.1711$; $u_1(1.0) \approx 5.5515$ and $u_2(1.0) \approx 13.3031$.

(c) Using a fourth-order Runge-Kutta method with $h = 0.1$, we get $u_1(0.5) \approx 2.1784$ and $u_2(0.5) \approx 4.7536$; $u_1(1.0) \approx 6.7731$ and $u_2(1.0) \approx 14.7205$.

5. (a) $(x(t), y(t), z(t)) = (0,5,0)$ for all the values of t specified. The particle doesn't seem to be moving.

(b) The functions x, y, and z seem to be increasing without bound as t grows larger, with the values of x, y, and z approaching each other.

7. Given $x(0) = 3$ and $y(0) = 2$, we can experiment with the values $t = 1, 2, 3$, and 4 to guess that the trajectory returns to its initial point for the first time when t is somewhere between 3 and 4. A finer examination reveals that t^* is between 3.7 and 3.8. Finally, by finding the values of x and y for $t = 3.71, 3.72, 3.73, \ldots$, and 3.8, we conclude that $t^* = 3.72$, to two decimal places.

9. (a) $\frac{dS}{dt} + \frac{dI}{dt} + \frac{dR}{dt} = \frac{d}{dt}(S + I + R) = 0$. This means that the total population does not change.

(b)

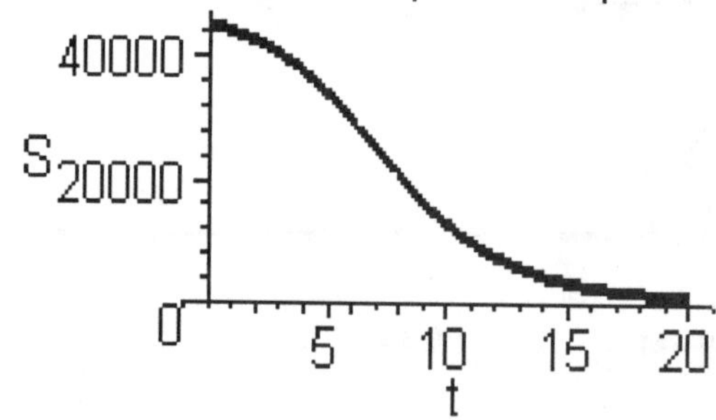

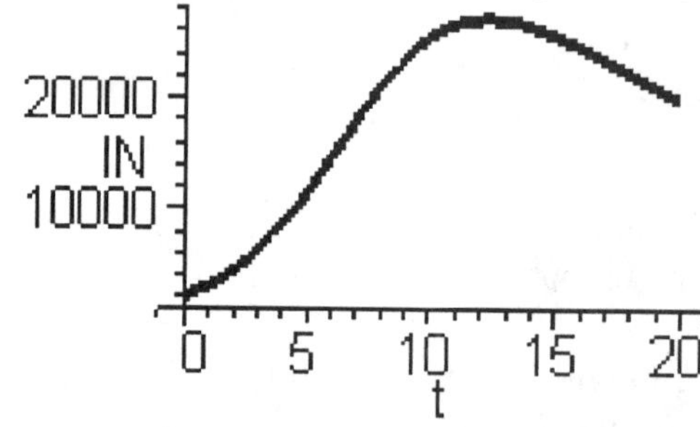

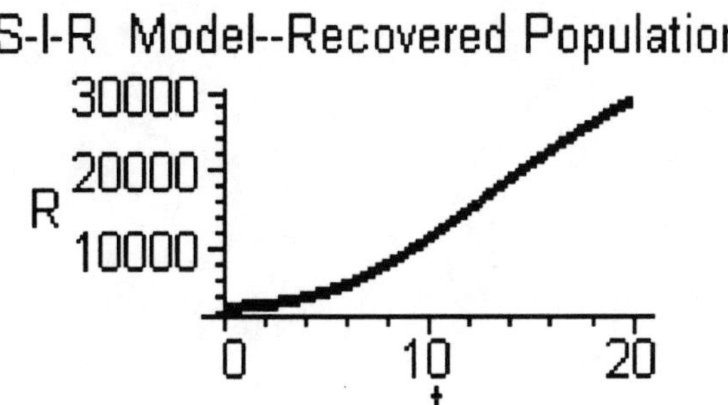

Here's a bonus—the graphs of S, I, and R against t on the same set of axes:

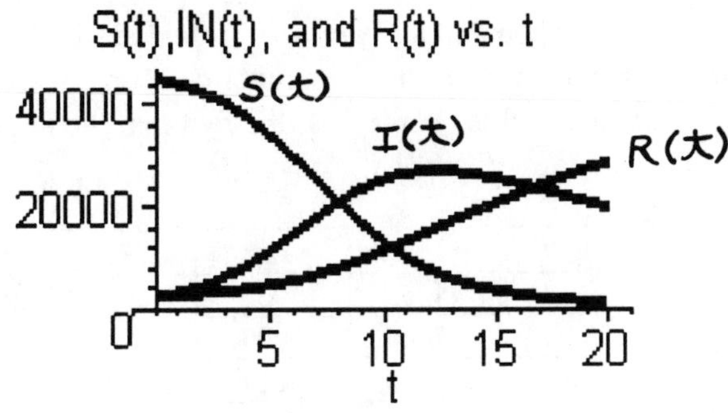

(c)

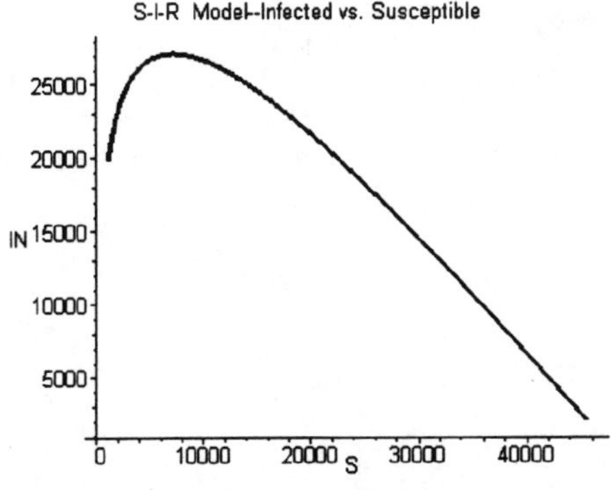

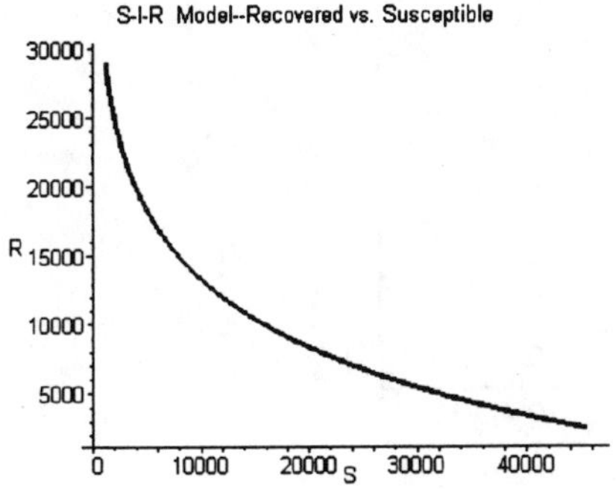

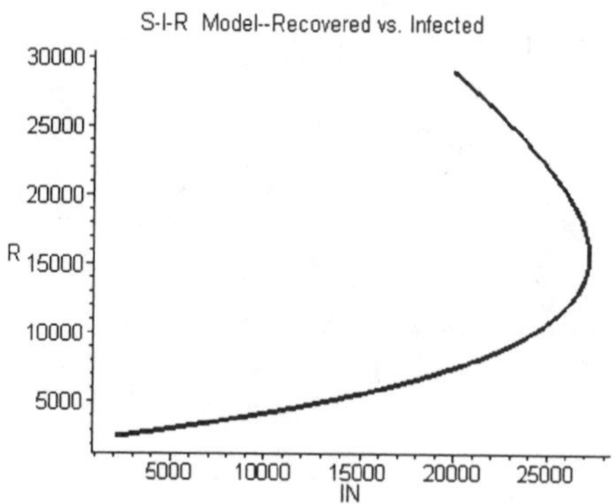

(d) We have used the *rkf45* method and all values are rounded to the nearest whole number. The values show the steady increase in the number of people who have recovered, the decreasing number of susceptible people, and the fact that the number of infected people probably peaks between days 10 and 15.

t	S	I	R
1	44255	3062	2682
2	42649	4405	2947
3	40460	6217	3323
10	13044	25547	11408
15	3447	25638	20915
16	2681	24609	22710
17	2108	23464	24428

(e) We conclude that $t \approx 161$ if we round down; but $t \approx 171$ if we round I to the nearest integer.

11. (a)

t	$x(t)$	$c(t)$
0.01	0.4492	-0.0158
0.02	0.4468	-0.0113
0.03	0.4432	-0.0068
0.04	0.4385	-0.0024
0.05	0.4330	0.0019
0.06	0.4266	0.0062
0.07	0.4196	0.0105
0.08	0.4120	0.0146
0.09	0.4039	0.0187
0.10	0.3952	0.0227

The direction of the solution curve is counterclockwise.

(b) The trajectory corresponding to the given initial conditions is

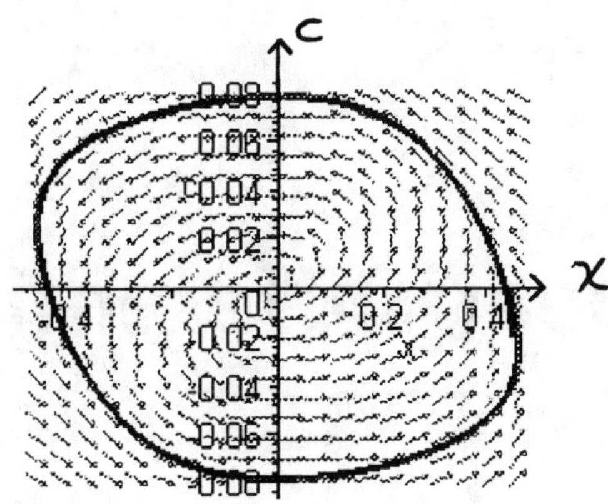

(c) Using *rkf45*, we find that $x(1.049) = 0.45996...$, $c(1.049) = -0.022555...$; $x(1.05) = 0.4598...$, $c(1.05) = -0.0220954...$; $x(1.051) = 0.4597437...$, $c(1.051) = -0.0216356.....$ We conclude that $t \approx 1.05$, or (rounded to the nearest tenth of a second) $t \approx 1.1$.

(d) Diastole: $(x, c) \approx (-0.46, 0.02)$ when $t \approx 0.52$; Systole: $(x, c) \approx (0.46, -0.02)$ when $t \approx 1.05$.

* **Project 4-1** is also given in Bugl's book of "explorations" [49]. This problem of the flow of lead pollution is treated briefly in [32] and more extensively in the text by Borrelli and Coleman [33; Sects. 7.1 and 7.10].*

Chapter 5

Systems of Linear Differential Equations

5.1 Systems and Matrices

1. (a) $\begin{pmatrix} 3 & 4 \\ -1 & -2 \end{pmatrix}\begin{pmatrix} x \\ y \end{pmatrix} = \begin{pmatrix} -7 \\ 5 \end{pmatrix}$; (b) $\begin{pmatrix} \pi & -3 \\ 5 & 2 \end{pmatrix}\begin{pmatrix} a \\ b \end{pmatrix} = \begin{pmatrix} 4 \\ -3 \end{pmatrix}$;

(c) $\begin{pmatrix} 1 & -1 & 1 \\ -1 & 2 & -3 \\ 2 & -3 & 5 \end{pmatrix}\begin{pmatrix} x \\ y \\ z \end{pmatrix} = \begin{pmatrix} 7 \\ 9 \\ 11 \end{pmatrix}$.

3. $\dot{X} = \begin{pmatrix} \dot{x} \\ \dot{y} \end{pmatrix} = \begin{pmatrix} 1 & -1 \\ -4 & 1 \end{pmatrix}\begin{pmatrix} x \\ y \end{pmatrix}$.

5. $\dot{X} = \begin{pmatrix} \dot{x} \\ \dot{y} \end{pmatrix} = \begin{pmatrix} 1 & 0 \\ 0 & 1 \end{pmatrix}\begin{pmatrix} x \\ y \end{pmatrix}$.

7. (a) Let $y_1 = y,\ y_2 = y'$. Then the system is $\{y_1' = y_2\,,\ y_2' = 3y_2 - 2y_1\}$, which can be written as $\begin{pmatrix} y_1' \\ y_2' \end{pmatrix} = \begin{pmatrix} 0 & 1 \\ -2 & 3 \end{pmatrix}\begin{pmatrix} y_1 \\ y_2 \end{pmatrix}$.

(b) Let $y_1 = y,\ y_2 = y'$. Then the system is $\{y_1' = y_2\,,\ y_2' = \frac{1}{5}y_1 - \frac{3}{5}y_2\}$, which can be written as $\begin{pmatrix} y_1' \\ y_2' \end{pmatrix} = \begin{pmatrix} 0 & 1 \\ \frac{1}{5} & -\frac{3}{5} \end{pmatrix}\begin{pmatrix} y_1 \\ y_2 \end{pmatrix}$.

9. We have $A(t)\,B(t) = \begin{pmatrix} a_{11}(t)b_{11}(t) + a_{12}(t)b_{21}(t) \\ a_{21}(t)b_{11}(t) + a_{22}(t)b_{21}(t) \end{pmatrix}$. Then

$$\frac{d}{dt}A(t)B(t) = \begin{pmatrix} \frac{d}{dt}\left(a_{11}(t)b_{11}(t) + a_{12}(t)b_{21}(t)\right) \\ \frac{d}{dt}\left(a_{21}(t)b_{11}(t) + a_{22}(t)b_{21}(t)\right) \end{pmatrix}$$

$$= \begin{pmatrix} a_{11}(t)\left(\frac{d}{dt}b_{11}(t)\right) + \left(\frac{d}{dt}a_{11}(t)\right)b_{11}(t) + a_{12}(t)\left(\frac{d}{dt}b_{21}(t)\right) + \left(\frac{d}{dt}a_{12}(t)\right)b_{21}(t) \\ a_{21}(t)\left(\frac{d}{dt}b_{11}(t)\right) + \left(\frac{d}{dt}a_{21}(t)\right)b_{11}(t) + a_{22}(t)\left(\frac{d}{dt}b_{21}(t)\right) + \left(\frac{d}{dt}a_{22}(t)\right)b_{21}(t) \end{pmatrix}$$

$$= \begin{pmatrix} a_{11}(t)\left(\frac{d}{dt}b_{11}(t)\right) + a_{12}(t)\left(\frac{d}{dt}b_{21}(t)\right) \\ a_{21}(t)\left(\frac{d}{dt}b_{11}(t)\right) + a_{22}(t)\left(\frac{d}{dt}b_{21}(t)\right) \end{pmatrix} + \begin{pmatrix} \left(\frac{d}{dt}a_{11}(t)\right)b_{11}(t) + \left(\frac{d}{dt}a_{12}(t)\right)b_{21}(t) \\ \left(\frac{d}{dt}a_{21}(t)\right)b_{11}(t) + \left(\frac{d}{dt}a_{22}(t)\right)b_{21}(t) \end{pmatrix}$$

$$= \begin{pmatrix} a_{11}(t) & a_{12}(t) \\ a_{21}(t) & a_{22}(t) \end{pmatrix} \begin{pmatrix} \frac{d}{dt}b_{11}(t) \\ \frac{d}{dt}b_{21}(t) \end{pmatrix} + \begin{pmatrix} \frac{d}{dt}a_{11}(t) & \frac{d}{dt}a_{12}(t) \\ \frac{d}{dt}a_{21}(t) & \frac{d}{dt}a_{21}(t) \end{pmatrix} \begin{pmatrix} b_{11}(t) \\ b_{21}(t) \end{pmatrix}$$

$$= A(t)\frac{d}{dt}B(t) + \frac{d}{dt}A(t)\,B(t).$$

11. We have the equation $\begin{pmatrix} 1 & 2 \\ 3 & 4 \end{pmatrix}\begin{pmatrix} x \\ y \end{pmatrix} = \begin{pmatrix} 1 \\ 3 \end{pmatrix}$, which is equivalent to the system

$$\begin{aligned} &(1): \; x + 2y = 1 \\ &(2): \; 3x + 4y = 3. \end{aligned}$$

If we multiply equation (1) by -2, we get the equivalent system

$$\begin{aligned} &(1)^*: \; -2x + -4y = -2 \\ &(2): \; 3x + 4y = 3. \end{aligned}$$

Adding the last two equations yields $x = 1$. Substitution of $x = 1$ into equation (1) yields $y = 0$. Therefore, $V = \begin{pmatrix} 1 \\ 0 \end{pmatrix}$.

5.2 Two-Dimensional Systems of First-Order Linear Equations

1. (a) $\begin{vmatrix} -3 & 5 \\ -4 & 1 \end{vmatrix} = (-3)(1) - 5(-4) = 17$

(b) $\begin{vmatrix} 4 & 2 \\ 10 & 5 \end{vmatrix} = (4)(5) - 2(10) = 0$

(c) $\begin{vmatrix} 6t & -4 \\ \sin t & t^3 \end{vmatrix} = (6t)(t^3) - (-4)(\sin t) = 6t^4 + 4\sin t$

(d) $\begin{vmatrix} \cos\theta & \sin\theta \\ -\sin\theta & \cos\theta \end{vmatrix} = (\cos\theta)(\cos\theta) - (\sin\theta)(-\sin\theta) = \cos^2\theta + \sin^2\theta = 1$

3. (a) $\begin{pmatrix} \dot{x} \\ \dot{y} \end{pmatrix} = \begin{pmatrix} 2 & 1 \\ 3 & 4 \end{pmatrix}\begin{pmatrix} x \\ y \end{pmatrix}$.

(b) The characteristic equation is $\lambda^2 - (1+1)\lambda + (1(1) - (-1)(-4)) = \lambda^2 - 6\lambda + 5 = 0$.

(c) $\lambda^2 - 6\lambda + 5 = (\lambda - 5)(\lambda - 1) = 0$, giving us $\lambda_1 = 5$ and $\lambda_2 = 1$.

(d) An eigenvector corresponding to an eigenvalue λ is a nonzero vector V such that $AV = \lambda V$. For $\lambda = 5$, we have $\begin{pmatrix} 2 & 1 \\ 3 & 4 \end{pmatrix}\begin{pmatrix} x \\ y \end{pmatrix} = 5\begin{pmatrix} x \\ y \end{pmatrix}$, which is equivalent to the system

$$\begin{matrix} 2x + y = 5x \\ 3x + 4y = 5y \end{matrix}, \text{ or } \begin{matrix} -3x + y = 0 \\ 3x - y = 0. \end{matrix}$$

The second equation is just the negative of the first equation, so that we really have only one equation, which we can write as $y = 3x$. Therefore, any nonzero vector of the form $\begin{pmatrix} x \\ 3x \end{pmatrix} = x\begin{pmatrix} 1 \\ 3 \end{pmatrix}$ is an eigenvector corresponding to the eigenvalue $\lambda = 5$. For $\lambda = 1$, we have $\begin{pmatrix} 2 & 1 \\ 3 & 4 \end{pmatrix}\begin{pmatrix} x \\ y \end{pmatrix} = (1)\begin{pmatrix} x \\ y \end{pmatrix} = \begin{pmatrix} x \\ y \end{pmatrix}$, which is equivalent to the system

$$\begin{matrix} 2x + y = x \\ 3x + 4y = y \end{matrix}, \text{ or } \begin{matrix} x + y = 0 \\ 3x + 3y = 0. \end{matrix}$$

Once again, we have only one equation, which we can write as $y = -x$. Therefore, any nonzero vector of the form $\begin{pmatrix} x \\ -x \end{pmatrix} = x\begin{pmatrix} 1 \\ -1 \end{pmatrix}$ is an eigenvector corresponding to the eigenvalue $\lambda = 1$.

5. (a) $\begin{pmatrix} \dot{x} \\ \dot{y} \end{pmatrix} = \begin{pmatrix} -4 & 2 \\ 2 & -1 \end{pmatrix} \begin{pmatrix} x \\ y \end{pmatrix}$.

(b) The characteristic equation is $\lambda^2 - (-4 + (-1))\lambda + ((-4)(-1) - 2(2)) = \lambda^2 + 5\lambda = 0$.

(c) $\lambda^2 + 5\lambda = \lambda(\lambda + 5) = 0$, so that $\lambda_1 = 0$ and $\lambda_2 = -5$.

(d) For $\lambda = 0$, we have $\begin{pmatrix} -4 & 2 \\ 2 & -1 \end{pmatrix} \begin{pmatrix} x \\ y \end{pmatrix} = (0) \begin{pmatrix} x \\ y \end{pmatrix} = \begin{pmatrix} 0 \\ 0 \end{pmatrix}$, which is equivalent to the system

$$\begin{aligned} -4x + 2y &= 0 \\ 2x - y &= 0 \end{aligned}.$$

The first equation is (-2) times the first equation, so that we really have only one equation, which we can write as $y = 2x$. Therefore, any nonzero vector of the form $\begin{pmatrix} x \\ 2x \end{pmatrix} = x\begin{pmatrix} 1 \\ 2 \end{pmatrix}$ is an eigenvector corresponding to the eigenvalue $\lambda = 0$. For $\lambda = -5$, we have $\begin{pmatrix} -4 & 2 \\ 2 & -1 \end{pmatrix} \begin{pmatrix} x \\ y \end{pmatrix} = (-5) \begin{pmatrix} x \\ y \end{pmatrix} = \begin{pmatrix} -5x \\ -5y \end{pmatrix}$, which is equivalent to the system

$$\begin{aligned} -4x + 2y &= -5x \\ 2x - y &= -5y \end{aligned}, \quad \text{or} \quad \begin{aligned} x + 2y &= 0 \\ 2x + 4y &= 0. \end{aligned}$$

Here we have only one equation, which we can write as $y = -x/2$. Therefore, any nonzero vector of the form $\begin{pmatrix} x \\ -x/2 \end{pmatrix} = x\begin{pmatrix} 1 \\ -1/2 \end{pmatrix} = x\begin{pmatrix} 2 \\ -1 \end{pmatrix}$ is an eigenvector corresponding to the eigenvalue $\lambda = -5$.

7. (a) $\begin{pmatrix} \dot{x} \\ \dot{y} \end{pmatrix} = \begin{pmatrix} -6 & 4 \\ -3 & 1 \end{pmatrix} \begin{pmatrix} x \\ y \end{pmatrix}$.

(b) The characteristic equation is $\lambda^2 - (-6 + 1)\lambda + ((-6)(1) - 4(-3)) = \lambda^2 + 5\lambda + 6 = 0$.

(c) $\lambda^2 + 5\lambda + 6 = (\lambda + 2)(\lambda + 3) = 0$, so that $\lambda_1 = -2$ and $\lambda_2 = -3$.

(d) For $\lambda = -2$, we have $\begin{pmatrix} \dot{x} \\ \dot{y} \end{pmatrix} = \begin{pmatrix} -6 & 4 \\ -3 & 1 \end{pmatrix} \begin{pmatrix} x \\ y \end{pmatrix} = (-2) \begin{pmatrix} x \\ y \end{pmatrix} = \begin{pmatrix} -2x \\ -2y \end{pmatrix}$, which is equivalent to the system

$$\begin{aligned} -6x + 4y &= -2x \\ -3x + y &= -2y \end{aligned}, \quad \text{or} \quad \begin{aligned} -4x + 4y &= 0 \\ -3x + 3y &= 0 \end{aligned}.$$

We really have only one equation here, which we can write as $y = x$. Therefore, any nonzero vector of the form $\begin{pmatrix} x \\ x \end{pmatrix} = x\begin{pmatrix} 1 \\ 1 \end{pmatrix}$ is an eigenvector corresponding to the eigenvalue $\lambda = -2$. For $\lambda = -3$, we have $\begin{pmatrix} \dot{x} \\ \dot{y} \end{pmatrix} = \begin{pmatrix} -6 & 4 \\ -3 & 1 \end{pmatrix}\begin{pmatrix} x \\ y \end{pmatrix} = (-3)\begin{pmatrix} x \\ y \end{pmatrix} = \begin{pmatrix} -3x \\ -3y \end{pmatrix}$, which is equivalent to the system

$$\begin{aligned} -6x + 4y &= -3x \\ -3x + y &= -3y \end{aligned}, \quad \text{or} \quad \begin{aligned} -3x + 4y &= 0 \\ -3x + 4y &= 0 \end{aligned}.$$

We have only one equation, which we can write as $y = 3x/4$. Therefore, any nonzero vector of the form $\begin{pmatrix} x \\ 3x/4 \end{pmatrix} = x\begin{pmatrix} 1 \\ 3/4 \end{pmatrix} = x\begin{pmatrix} 4 \\ 3 \end{pmatrix}$ is an eigenvector corresponding to the eigenvalue $\lambda = -3$.

9. (a) Substituting for $x(t)$ and $y(t)$ in the right-hand side of the expression $\frac{dy}{dx} = \frac{-4x + 3y}{-2x + y}$, we get $\frac{8C_1e^{2t} - C_2e^{-t}}{2C_1e^{2t} - C_2e^{-t}}$.

(b) We can write $\frac{8C_1e^{2t} - C_2e^{-t}}{2C_1e^{2t} - C_2e^{-t}} = \frac{e^{2t}(8C_1 - C_2e^{-3t})}{e^{2t}(2C_1 - C_2e^{-3t})} = \frac{8C_1 - C_2e^{-3t}}{2C_1 - C_2e^{-3t}}$ and see that $\frac{8C_1 - C_2e^{-3t}}{2C_1 - C_2e^{-3t}} \to \frac{8C_1 - 0}{2C_1 - 0} = 4$ as $t \to \infty$ provided that $C_1 \neq 0$—that is, provided that the trajectory is not on the line $\begin{pmatrix} C_2e^{-t} \\ C_2e^{-t} \end{pmatrix} = C_2e^{-t}\begin{pmatrix} 1 \\ 1 \end{pmatrix}$ determined by the representative eigenvector $\begin{pmatrix} 1 \\ 1 \end{pmatrix}$.

(c) Using the result of part (a), we can write $\frac{8C_1e^{2t} - C_2e^{-t}}{2C_1e^{2t} - C_2e^{-t}} = \frac{e^{-t}(8C_1e^{3t} - C_2)}{e^{-t}(2C_1e^{3t} - C_2)}$ $= \frac{8C_1e^{3t} - C_2}{2C_1e^{3t} - C_2} \to \frac{0 - C_2}{0 - C_2} = 1$ as $t \to -\infty$ provided that $C_2 \neq 0$—that is, provided that the trajectory is not on the line $\begin{pmatrix} C_1e^{2t} \\ 4C_1e^{2t} \end{pmatrix} = C_1e^{2t}\begin{pmatrix} 1 \\ 4 \end{pmatrix}$ determined by the representative eigenvector $\begin{pmatrix} 1 \\ 4 \end{pmatrix}$.

11. The phase portrait corresponding to the system in Exercise 4 is

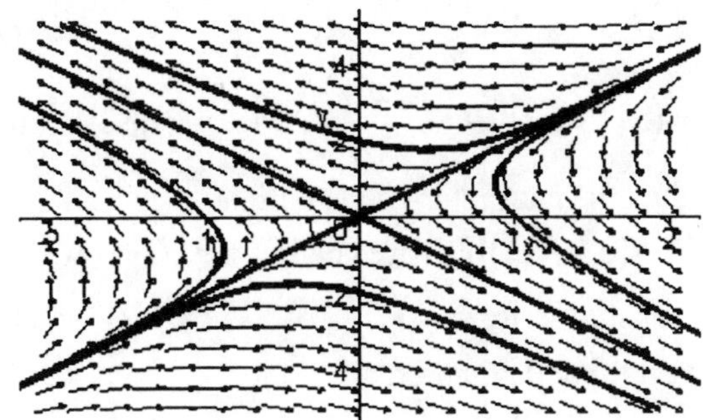

The trajectories are moving away from the origin as t increases. Algebraically, this is a consequence of the fact that one eigenvalue, 3, is positive. Furthermore the trajectories approach the line determined by the representative eigenvector $\begin{pmatrix} 1 \\ -2 \end{pmatrix}$ associated with the larger eigenvalue 3 as $t \to \infty$. The trajectories approach the line determined by $\begin{pmatrix} 1 \\ 2 \end{pmatrix}$, the representative eigenvector associated with the smaller eigenvalue -1 as $t \to -\infty$.

13. (a) First we write the system in the form $\begin{pmatrix} \frac{dc_1}{dx} \\ \frac{dc_2}{dx} \end{pmatrix} = \begin{pmatrix} -\alpha_1 & \alpha_1 \\ -\alpha_2 & \alpha_2 \end{pmatrix} \begin{pmatrix} c_1 \\ c_2 \end{pmatrix}$. Then we use a computer algebra system to find the eigenvalues and corresponding eigenvectors:

$\lambda_1 = 0,\ \lambda_2 = \alpha_2 - \alpha_1$; $V_1 = \begin{pmatrix} 1 \\ 1 \end{pmatrix}, V_2 = \begin{pmatrix} \alpha_1/\alpha_2 \\ 1 \end{pmatrix} = \begin{pmatrix} \alpha_1 \\ \alpha_2 \end{pmatrix}$. The general solution is

$F(x) = \begin{pmatrix} c_1(x) \\ c_2(x) \end{pmatrix} = K_1 \begin{pmatrix} 1 \\ 1 \end{pmatrix} e^{0 \cdot x} + K_2 \begin{pmatrix} \alpha_1 \\ \alpha_2 \end{pmatrix} e^{(\alpha_2 - \alpha_1)x} = \begin{pmatrix} K_1 + K_2 \alpha_1 e^{(\alpha_2 - \alpha_1)x} \\ K_1 + K_2 \alpha_2 e^{(\alpha_2 - \alpha_1)x} \end{pmatrix}$. The initial conditions yield the algebraic system

$$c_0 = K_1 + K_2 \alpha_1$$
$$C_0 = K_1 + K_2 \alpha_2$$

with solution $K_1 = \dfrac{\alpha_2 c_0 - \alpha_1 C_0}{\alpha_2 - \alpha_1}$ and $K_2 = \dfrac{C_0 - c_0}{\alpha_2 - \alpha_1}$. Therefore we have

$$c_1(x) = \alpha_1 \left(\frac{C_0 - c_0}{\alpha_2 - \alpha_1} \right) e^{(\alpha_2 - \alpha_1)x} + \frac{\alpha_2 c_0 - \alpha_1 C_0}{\alpha_2 - \alpha_1} \text{ and}$$

$$c_2(x) = \alpha_2 \left(\frac{C_0 - c_0}{\alpha_2 - \alpha_1} \right) e^{(\alpha_2 - \alpha_1)x} + \frac{\alpha_2 c_0 - \alpha_1 C_0}{\alpha_2 - \alpha_1}.$$

(b) For example, you could derive the single equation $\frac{d^2c_1}{dx^2}+(\alpha_1-\alpha_2)\frac{dc_1}{dx}=0$:

$$\frac{d^2c_1}{dx^2}=\frac{d}{dx}[-\alpha_1(c_1-c_2)]=-\alpha_1\frac{dc_1}{dx}+\alpha_1\frac{dc_2}{dx}=-\alpha_1\frac{dc_1}{dx}+\alpha_1\left[\frac{\alpha_2}{\alpha_1}\frac{dc_1}{dx}\right]$$

$=-\alpha_1\frac{dc_1}{dx}+\alpha_2\frac{dc_1}{dx}=(\alpha_2-\alpha_1)\frac{dc_1}{dx}$, or $\frac{d^2c_1}{dx^2}+(\alpha_1-\alpha_2)\frac{dc_1}{dx}=0$. Now the differential equation has the characteristic equation $\lambda^2+(\alpha_1-\alpha_2)\lambda=\lambda(\lambda+(\alpha_1-\alpha_2))=0$, which has the roots $\lambda_1=0$ and $\lambda_2=\alpha_2-\alpha_1$. Therefore, the general solution is $c_1(x)=K_1e^{0\cdot x}+K_2e^{(\alpha_2-\alpha_1)x}=K_1+K_2e^{(\alpha_2-\alpha_1)x}$. The initial conditions for this single equation are

$c_1(0)=c_0$ and $\frac{dc_1}{dx}(0)=-\alpha_1(c_1(0)-c_2(0))=-\alpha_1(c_0-C_0)$, leading

to the conclusion that $K_1=\frac{\alpha_2c_0-\alpha_1C_0}{\alpha_2-\alpha_1}$ and $K_2=\frac{C_0-c_0}{\alpha_2-\alpha_1}$. Then we have

$c_1(x)=\alpha_1\left(\frac{C_0-c_0}{\alpha_2-\alpha_1}\right)e^{(\alpha_2-\alpha_1)x}+\frac{\alpha_2c_0-\alpha_1C_0}{\alpha_2-\alpha_1}$ as in part (a). In a similar way, we could have derived a single second-order equation involving c_2 and solved it via the technique shown in Section 4.1. More directly, because we have already solved for $c_1(x)$, we can use the first equation in our system to see that $c_2(x)=\frac{1}{\alpha_1}\left[\frac{dc_1}{dx}+\alpha_1c_1\right]=\frac{1}{\alpha_1}\frac{dc_1}{dx}+c_1$.

5.3 Stability of Linear Systems: Unequal Real Eigenvalues

1. (a) The system can be expressed as $\begin{pmatrix}\dot{x}\\ \dot{y}\end{pmatrix}=\begin{pmatrix}3&0\\0&2\end{pmatrix}$. The characteristic equation is $\lambda^2-(3+2)\lambda+(3(2)-0)=\lambda^2-5\lambda+6=(\lambda-3)(\lambda-2)=0$, so that the eigenvalues are $\lambda_1=3$ and $\lambda_2=2$. For $\lambda=3$, we have $\begin{pmatrix}3&0\\0&2\end{pmatrix}\begin{pmatrix}x\\y\end{pmatrix}=3\begin{pmatrix}x\\y\end{pmatrix}=\begin{pmatrix}3x\\3y\end{pmatrix}$, which is equivalent to the system

$$\begin{aligned}3x&=3x\\2y&=3y\end{aligned}$$

This system implies that x is arbitrary and y must be zero. Therefore, any nonzero vector of the form $\begin{pmatrix}x\\0\end{pmatrix}=x\begin{pmatrix}1\\0\end{pmatrix}$ is an eigenvector corresponding to the eigenvalue $\lambda=3$. For $\lambda=2$, we have $\begin{pmatrix}3&0\\0&2\end{pmatrix}\begin{pmatrix}x\\y\end{pmatrix}=2\begin{pmatrix}x\\y\end{pmatrix}=\begin{pmatrix}2x\\2y\end{pmatrix}$, which is equivalent to the system

$$\begin{aligned} 3x &= 2x \\ 2y &= 2y \end{aligned},$$

implying that $x = 0$ and y is arbitrary. Therefore, any nonzero vector of the form $\begin{pmatrix} 0 \\ y \end{pmatrix} = y\begin{pmatrix} 0 \\ 1 \end{pmatrix}$ is an eigenvector corresponding to the eigenvalue $\lambda = 2$.

(b) Here is a plot of several trajectories with the eigenvectors (essentially, the x- and y-axes) shown:

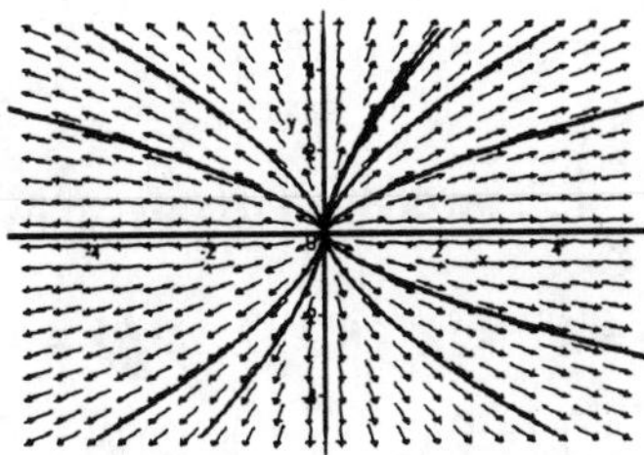

3. (a) The system is $\begin{pmatrix} x' \\ y' \end{pmatrix} = \begin{pmatrix} -3 & -1 \\ 4 & 2 \end{pmatrix}\begin{pmatrix} x \\ y \end{pmatrix}$. The characteristic equation is $\lambda^2 - (-3+2)\lambda + ((-3)(2)-(-1)(4)) = \lambda^2 + \lambda - 2 = (\lambda - 1)(\lambda + 2) = 0$, so that $\lambda_1 = 1$ and $\lambda_2 = -2$. For $\lambda = 1$, we have $\begin{pmatrix} x' \\ y' \end{pmatrix} = \begin{pmatrix} -3 & -1 \\ 4 & 2 \end{pmatrix}\begin{pmatrix} x \\ y \end{pmatrix} = (1)\begin{pmatrix} x \\ y \end{pmatrix} = \begin{pmatrix} x \\ y \end{pmatrix}$, which is equivalent to the system

$$\begin{aligned} -3x - y &= x \\ 4x + 2y &= y \end{aligned}, \text{ or } \begin{aligned} -4x - y &= 0 \\ 4x + y &= 0. \end{aligned}$$

We have only one equation, which can be written as $y = -4x$. Therefore, any nonzero vector of the form $\begin{pmatrix} x \\ -4x \end{pmatrix} = x\begin{pmatrix} 1 \\ -4 \end{pmatrix}$ is an eigenvector corresponding to the eigenvalue $\lambda = 1$. For $\lambda = -2$, we have $\begin{pmatrix} x' \\ y' \end{pmatrix} = \begin{pmatrix} -3 & -1 \\ 4 & 2 \end{pmatrix}\begin{pmatrix} x \\ y \end{pmatrix} = (-2)\begin{pmatrix} x \\ y \end{pmatrix} = \begin{pmatrix} -2x \\ -2y \end{pmatrix}$, which is equivalent to the system

$$\begin{aligned} -3x - y &= -2x \\ 4x + 2y &= -2y \end{aligned}, \text{ or } \begin{aligned} -x - y &= 0 \\ 4x + 4y &= 0. \end{aligned}$$

There is only one distinct equation, which can be written as $y = -x$. Therefore, any nonzero vector of the form $\begin{pmatrix} x \\ -x \end{pmatrix} = x\begin{pmatrix} 1 \\ -1 \end{pmatrix}$ is an eigenvector corresponding to the eigenvalue $\lambda = -2$.

(b) Here's the plot of several trajectories and the eigenvectors:

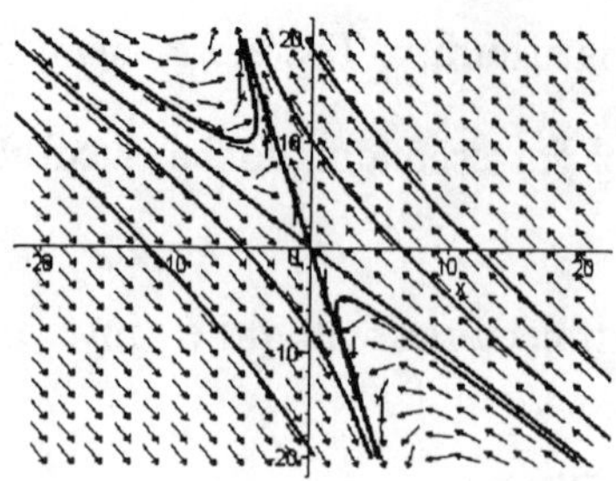

5. (a) The system is $\begin{pmatrix}\dot{x}\\ \dot{y}\end{pmatrix}=\begin{pmatrix}1 & 5\\ 1 & -3\end{pmatrix}\begin{pmatrix}x\\ y\end{pmatrix}$. The characteristic equation is $\lambda^2-(1+(-3))\lambda+(1(-3)-5(1))=\lambda^2+2\lambda-8=(\lambda-2)(\lambda+4)=0$, so that $\lambda_1=2$ and $\lambda_2=-4$. For $\lambda=2$, we have $\begin{pmatrix}\dot{x}\\ \dot{y}\end{pmatrix}=\begin{pmatrix}1 & 5\\ 1 & -3\end{pmatrix}\begin{pmatrix}x\\ y\end{pmatrix}=2\begin{pmatrix}x\\ y\end{pmatrix}=\begin{pmatrix}2x\\ 2y\end{pmatrix}$, which is equivalent to the system

$$\begin{matrix}x+5y=2x\\ x-3y=2y\end{matrix}, \text{ or } \begin{matrix}-x+5y=0\\ x-5y=0.\end{matrix}$$

We have only one equation, which can be written as $y=\frac{1}{5}x$. Therefore, any nonzero vector of the form $\begin{pmatrix}x\\ \frac{1}{5}x\end{pmatrix}=x\begin{pmatrix}1\\ \frac{1}{5}\end{pmatrix}=x\begin{pmatrix}5\\ 1\end{pmatrix}$ is an eigenvector corresponding to the eigenvalue $\lambda=2$. For $\lambda=-4$, we have $\begin{pmatrix}\dot{x}\\ \dot{y}\end{pmatrix}=\begin{pmatrix}1 & 5\\ 1 & -3\end{pmatrix}\begin{pmatrix}x\\ y\end{pmatrix}=(-4)\begin{pmatrix}x\\ y\end{pmatrix}=\begin{pmatrix}-4x\\ -4y\end{pmatrix}$, which is equivalent to the system

$$\begin{matrix}x+5y=-4x\\ x-3y=-4y\end{matrix}, \text{ or } \begin{matrix}5x+5y=0\\ x+y=0.\end{matrix}$$

There is only one distinct equation, which can be written as $y=-x$. Therefore, any nonzero vector of the form $\begin{pmatrix}x\\ -x\end{pmatrix}=x\begin{pmatrix}1\\ -1\end{pmatrix}$ is an eigenvector corresponding to the eigenvalue $\lambda=-4$.

(b) Trajectories and eigenvectors are shown below:

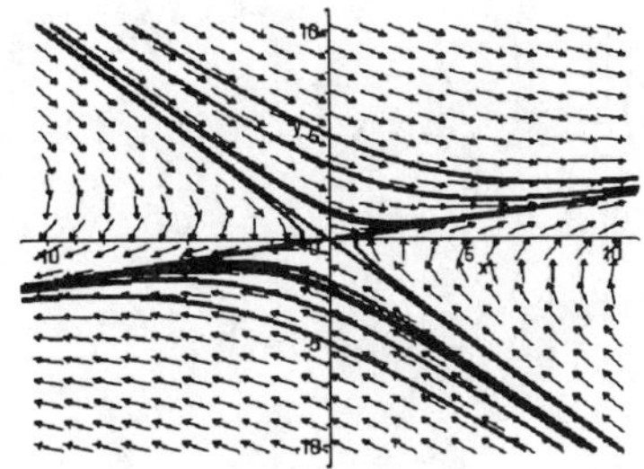

7. (a) The system can be written as $\begin{pmatrix} \dot{x} \\ \dot{y} \end{pmatrix} = \begin{pmatrix} -3 & 1 \\ 4 & -2 \end{pmatrix}\begin{pmatrix} x \\ y \end{pmatrix}$. The characteristic equation is $\lambda^2 - (-3 + (-2))\lambda + ((-3)(-2) - 1(4)) = \lambda^2 + 5\lambda + 2 = 0$. The quadratic formula yields $\lambda_1 = \frac{-5+\sqrt{17}}{2}$ and $\lambda_2 = \frac{-5-\sqrt{17}}{2}$, both irrational numbers. Using a CAS, we find corresponding representative eigenvectors $V_1 = \begin{pmatrix} \frac{-1+\sqrt{17}}{8} \\ 1 \end{pmatrix}$ and $V_2 = \begin{pmatrix} \frac{-1-\sqrt{17}}{8} \\ 1 \end{pmatrix}$.

(b) Here are some trajectories and the eigenvectors (which are difficult to pick out):

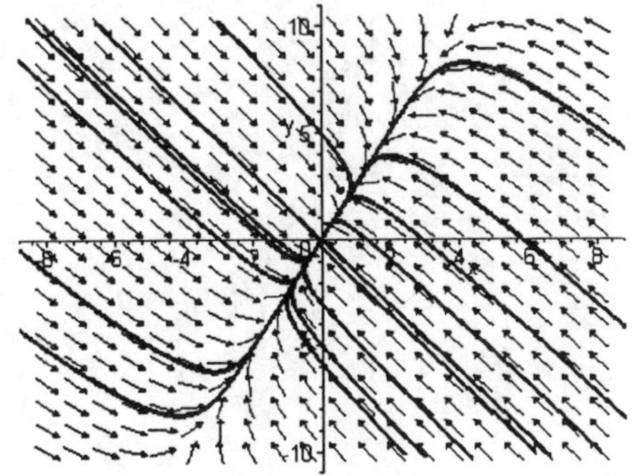

9. (a) The system is $\begin{pmatrix} x' \\ y' \end{pmatrix} = \begin{pmatrix} -2 & -1 \\ -1 & 2 \end{pmatrix}\begin{pmatrix} x \\ y \end{pmatrix}$. The characteristic equation is $\lambda^2 - (-2+2)\lambda + ((-2)(2) - (-1)(-1)) = \lambda^2 - 5 =$, so that $\lambda_1 = \sqrt{5}$ and $\lambda_2 = -\sqrt{5}$. For $\lambda = \sqrt{5}$, we have $\begin{pmatrix} x' \\ y' \end{pmatrix} = \begin{pmatrix} -2 & -1 \\ -1 & 2 \end{pmatrix}\begin{pmatrix} x \\ y \end{pmatrix} = \sqrt{5}\begin{pmatrix} x \\ y \end{pmatrix} = \begin{pmatrix} \sqrt{5}\,x \\ \sqrt{5}\,x \end{pmatrix}$, which is equivalent to the system

$$\begin{aligned} -2x - \ \ y &= \sqrt{5}\,x \\ -x + 2y &= \sqrt{5}\,y \end{aligned}, \quad \text{or} \quad \begin{aligned} -\left(2+\sqrt{5}\right)x - \ \ y &= 0 \\ -x + \left(2-\sqrt{5}\right)y &= 0. \end{aligned}$$

We have only one equation, which can be written as $y = -\left(2+\sqrt{5}\right)x$. Therefore, any nonzero vector of the form $\begin{pmatrix} x \\ -(2+\sqrt{5}) \end{pmatrix} = x\begin{pmatrix} 1 \\ -(2+\sqrt{5}) \end{pmatrix} = x\begin{pmatrix} 2-\sqrt{5} \\ 1 \end{pmatrix}$ is an eigenvector corresponding to the eigenvalue $\lambda = \sqrt{5}$. For $\lambda = -\sqrt{5}$, we have $\begin{pmatrix} x' \\ y' \end{pmatrix} = \begin{pmatrix} -2 & -1 \\ -1 & 2 \end{pmatrix}\begin{pmatrix} x \\ y \end{pmatrix} = -\sqrt{5}\begin{pmatrix} x \\ y \end{pmatrix} = \begin{pmatrix} -\sqrt{5}\,x \\ -\sqrt{5}\,x \end{pmatrix}$, which is equivalent to the system

$$\begin{aligned} -2x - \ \ y &= -\sqrt{5}\,x \\ -x + 2y &= -\sqrt{5}\,y \end{aligned}, \quad \text{or} \quad \begin{aligned} -\left(2-\sqrt{5}\right)x - \ \ y &= 0 \\ -x + \left(2+\sqrt{5}\right)y &= 0. \end{aligned}$$

There is only one distinct equation, which can be written as $y = -\left(2-\sqrt{5}\right)x$. Therefore, any nonzero vector of the form $\begin{pmatrix} x \\ -\left(2-\sqrt{5}\right)x \end{pmatrix} = x\begin{pmatrix} 1 \\ -\left(2-\sqrt{5}\right) \end{pmatrix} = x\begin{pmatrix} 2+\sqrt{5} \\ 1 \end{pmatrix}$ is an eigenvector corresponding to the eigenvalue $\lambda = -\sqrt{5}$.

(b) Here are some trajectories and the eigenvectors:

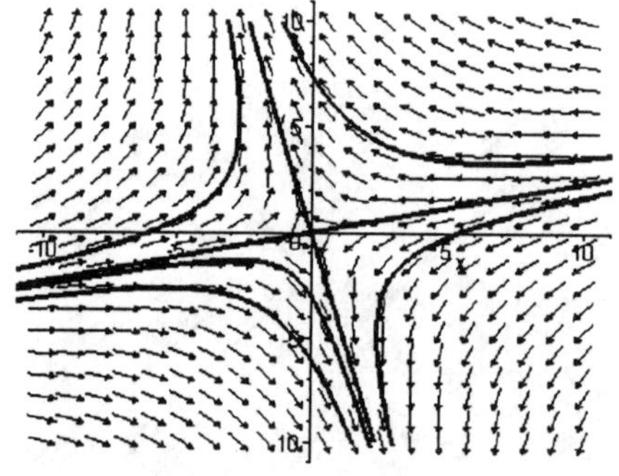

11. (a) We can write the system as $\begin{pmatrix} \dot{x} \\ \dot{y} \end{pmatrix} = \begin{pmatrix} 4 & -3 \\ 8 & -6 \end{pmatrix}\begin{pmatrix} x \\ y \end{pmatrix}$. The characteristic equation is $\lambda^2 - (4+(-6))\lambda + (4(-6)-(-3)8) = \lambda^2 + 2\lambda = \lambda(\lambda+2) = 0$, so that the eigenvalues of this system are $\lambda = 0$ and $\lambda = -2$.

(b) For $\lambda = 0$, we must have $\begin{pmatrix} \dot{x} \\ \dot{y} \end{pmatrix} = \begin{pmatrix} 4 & -3 \\ 8 & -6 \end{pmatrix}\begin{pmatrix} x \\ y \end{pmatrix} = (0)\begin{pmatrix} x \\ y \end{pmatrix} = \begin{pmatrix} 0 \\ 0 \end{pmatrix}$, which is equivalent to the system

$$\begin{aligned} 4x - 3y &= 0 \\ 8x - 6y &= 0, \end{aligned}$$

or $y = \frac{4}{3}x$. This means that the eigenvectors corresponding to $\lambda = 0$ must have the form

$\begin{pmatrix} x \\ \frac{4}{3}x \end{pmatrix} = x\begin{pmatrix} 1 \\ \frac{4}{3} \end{pmatrix} = x\begin{pmatrix} 3 \\ 4 \end{pmatrix}$, $x \neq 0$. For $\lambda = -2$, we get $\begin{pmatrix} \dot{x} \\ \dot{y} \end{pmatrix} = \begin{pmatrix} 4 & -3 \\ 8 & -6 \end{pmatrix}\begin{pmatrix} x \\ y \end{pmatrix} = (-2)\begin{pmatrix} x \\ y \end{pmatrix} = \begin{pmatrix} -2x \\ -2y \end{pmatrix}$,

which is equivalent to the system

$$\begin{matrix} 4x - 3y = -2x \\ 8x - 6y = -2y \end{matrix}, \quad \text{or} \quad \begin{matrix} 6x - 3y = 0 \\ 8x - 4y = 0. \end{matrix}$$

We have the single equation $y = 2x$, so that any eigenvector corresponding to the eigenvalue $\lambda = -2$ must have the form $\begin{pmatrix} x \\ 2x \end{pmatrix} = x\begin{pmatrix} 1 \\ 2 \end{pmatrix}$, $x \neq 0$. Therefore representative eigenvectors corresponding to the eigenvalues $\lambda_1 = 0$ and $\lambda_2 = -2$ are $V_1 = \begin{pmatrix} 3 \\ 4 \end{pmatrix}$ and $V_2 = \begin{pmatrix} 1 \\ 2 \end{pmatrix}$, respectively.

(c) Here's a plot of some trajectories:

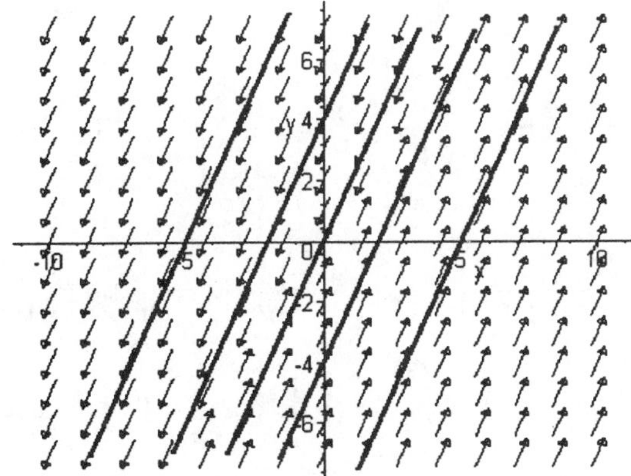

Every point of the line $y = \frac{4}{3}x$ is an equilibrium point. The origin is a *sink*, while every other point on the line is a *node*. All other trajectories (straight lines) seem to be parallel to the trajectory determined by the eigenvector V_1.

(d) We can write the general solution of the system in the form given by equation (5.3.1) in the text: $X(t) = c_1 e^{0 \cdot t}\begin{pmatrix} 3 \\ 4 \end{pmatrix} + c_2 e^{-2t}\begin{pmatrix} 1 \\ 2 \end{pmatrix} = c_1\begin{pmatrix} 3 \\ 4 \end{pmatrix} + c_2 e^{-2t}\begin{pmatrix} 1 \\ 2 \end{pmatrix}$. This shows that every trajectory is indeed parallel to V_1. Furthermore, the trajectories, the points on the lines viewed as connected to the origin, approach the eigenvector V_1 as $t \to \infty$. (See the explanation just before Example 5.3.4 in the text.)

13. The idea in this problem is to construct the characteristic polynomial and then use the fact that the coefficient of λ is the negative of the trace, $a + d$, of the matrix of coefficients $A = \begin{pmatrix} a & b \\ c & d \end{pmatrix}$, whereas the constant term is the determinant, $ad - bc$, of A.

(a) The characteristic polynomial is $(\lambda-(-3))(\lambda-(-5))=(\lambda+3)(\lambda+5)=\lambda^2+8\lambda+15$, so that tr($A$) = -8 and det(A) = 15. An appropriate matrix, for example, is $\begin{pmatrix}-4 & 1\\ 1 & -4\end{pmatrix}$, with resulting system $\{\dot{x}=-4x+y,\ \dot{y}=x-4y\}$.

(b) The characteristic polynomial is $(\lambda-(-1))(\lambda-4)=(\lambda+1)(\lambda-4)=\lambda^2-3\lambda-4$, implying that tr($A$) = 3 and det($A$) = -4. One such matrix is $\begin{pmatrix}1 & 2\\ 3 & 2\end{pmatrix}$, with resulting system $\{\dot{x}=x+2y,\ \dot{y}=3x+2y\}$.

(c) The characteristic polynomial is $(\lambda-2)(\lambda-3)=\lambda^2-5\lambda+6$, implying that tr($A$) = 5 and det($A$) = 6. One such matrix is $\begin{pmatrix}2 & 0\\ 0 & 3\end{pmatrix}$, with resulting system $\{\dot{x}=2x,\ \dot{y}=3y$.

15. (a) We write the system in the form $\begin{pmatrix}\dot{x}\\ \dot{y}\end{pmatrix}=\begin{pmatrix}-\frac{P}{V_1} & \frac{P}{V_2}\\ \frac{P}{V_1} & -\frac{P}{V_2}\end{pmatrix}\begin{pmatrix}x\\ y\end{pmatrix}$. The characteristic equation is

$\lambda^2-\left(-\frac{P}{V_1}-\frac{P}{V_2}\right)\lambda+\left(\frac{P^2}{V_1V_2}-\frac{P^2}{V_1V_2}\right)=\lambda^2+\left(\frac{P}{V_1}+\frac{P}{V_2}\right)\lambda=\lambda\left(\lambda+P\left(\frac{1}{V_1}+\frac{1}{V_2}\right)\right)=0$, so that the eigenvalues are $\lambda_1=0$ and $\lambda_2=-P\left(\frac{1}{V_1}+\frac{1}{V_2}\right)$. For $\lambda=0$, we have

$\begin{pmatrix}\dot{x}\\ \dot{y}\end{pmatrix}=\begin{pmatrix}-\frac{P}{V_1} & \frac{P}{V_2}\\ \frac{P}{V_1} & -\frac{P}{V_2}\end{pmatrix}\begin{pmatrix}x\\ y\end{pmatrix}=(0)\begin{pmatrix}x\\ y\end{pmatrix}=\begin{pmatrix}0\\ 0\end{pmatrix}$, which is equivalent to the system

$$\begin{aligned}(-P/V_1)x+(P/V_2)y&=0\\ (P/V_1)x-(P/V_2)y&=0.\end{aligned}$$

We have only one equation, which can be written as $y=(V_2/V_1)x$. Therefore, any nonzero vector of the form $\begin{pmatrix}x\\ (V_2/V_1)x\end{pmatrix}=x\begin{pmatrix}1\\ V_2/V_1\end{pmatrix}=x\begin{pmatrix}V_1\\ V_2\end{pmatrix}$ is an eigenvector corresponding to the eigenvalue $\lambda=0$. For $\lambda=-P\left(\frac{1}{V_1}+\frac{1}{V_2}\right)$, we have

$$\begin{pmatrix}\dot{x}\\ \dot{y}\end{pmatrix}=\begin{pmatrix}-\dfrac{P}{V_1} & \dfrac{P}{V_2}\\ \dfrac{P}{V_1} & -\dfrac{P}{V_2}\end{pmatrix}\begin{pmatrix}x\\ y\end{pmatrix}=-P\left(\frac{1}{V_1}+\frac{1}{V_2}\right)\begin{pmatrix}x\\ y\end{pmatrix}$$, which is

equivalent to the system

$$\begin{aligned}(-P/V_1)x+(P/V_2)y&=-P(1/V_1+1/V_2)x\\ (P/V_1)x-(P/V_2)y&=-P(1/V_1+1/V_2)y\end{aligned}, \quad\text{or}\quad \begin{aligned}(P/V_2)y&=-(P/V_2)x\\ (P/V_1)x&=-(P/V_1)y.\end{aligned}$$

There is only one distinct equation, which can be written as $y=-x$. Therefore, any nonzero vector of the form $\begin{pmatrix}x\\ -x\end{pmatrix}=x\begin{pmatrix}1\\ -1\end{pmatrix}$ is an eigenvector corresponding to the eigenvalue $\lambda=-P\left(\frac{1}{V_1}+\frac{1}{V_2}\right)$. Then we can write the general solution of the system as $X(t)=c_1\begin{pmatrix}V_1\\ V_2\end{pmatrix}+c_2e^{-P(1/V_1+1/V_2)t}\begin{pmatrix}1\\ -1\end{pmatrix}$, so that $X(0)=c_1\begin{pmatrix}V_1\\ V_2\end{pmatrix}+c_2\begin{pmatrix}1\\ -1\end{pmatrix}=\begin{pmatrix}x_0\\ y_0\end{pmatrix}$. This is equivalent to the system

$$\begin{aligned}V_1c_1+c_2&=x_0\\ V_2c_1-c_2&=y_0,\end{aligned}$$

whose solution is $c_1=\dfrac{x_0+y_0}{V_1+V_2}$ and $c_2=\dfrac{x_0V_2-y_0V_1}{V_1+V_2}$. Now the solution of the IVP can be written in the form $X(t)=\left(\dfrac{x_0+y_0}{V_1+V_2}\right)\begin{pmatrix}V_1\\ V_2\end{pmatrix}+\left(\dfrac{x_0V_2-y_0V_1}{V_1+V_2}\right)e^{-P(1/V_1+1/V_2)t}\begin{pmatrix}1\\ -1\end{pmatrix}$, or

$$x(t)=\left(\frac{x_0+y_0}{V_1+V_2}\right)V_1+\left(\frac{x_0V_2-y_0V_1}{V_1+V_2}\right)e^{-P\left(\frac{1}{V_1}+\frac{1}{V_2}\right)t}\quad\text{and}$$

$$y(t)=\left(\frac{x_0+y_0}{V_1+V_2}\right)V_2-\left(\frac{x_0V_2-y_0V_1}{V_1+V_2}\right)e^{-P\left(\frac{1}{V_1}+\frac{1}{V_2}\right)t}.$$

(b) Using the results of part (a). it is easy to see that

$$\lim_{t\to\infty}x(t)=\left(\frac{x_0+y_0}{V_1+V_2}\right)V_1\quad\text{and}\quad\lim_{t\to\infty}y(t)=\left(\frac{x_0+y_0}{V_1+V_2}\right)V_2.$$

(c) $\lim_{t\to\infty}[x(t)+y(t)]=\left(\dfrac{x_0+y_0}{V_1+V_2}\right)V_1+\left(\dfrac{x_0+y_0}{V_1+V_2}\right)V_2=x_0+y_0$. Physically, this says that after a long time, the total quantity of solution on both sides of the membrane still equals the total quantity present originally—that is, the total at $t=0$. Nothing is gained, nothing is lost.

(d) The chemical moves across the membrane from the side with a higher concentration to the side with a lower concentration: If $\frac{y}{V_2} > \frac{x}{V_1}$, then $\dot{x} = P\left(\frac{y}{V_2} - \frac{x}{V_1}\right) > 0$, which means that x, the amount of chemical with the lower concentration, is increasing; whereas if $\frac{x}{V_1} > \frac{y}{V_2}$, then $\dot{y} = P\left(\frac{x}{V_1} - \frac{y}{V_2}\right) > 0$, meaning that y, now the amount of chemical with the lower concentration, is increasing.

17. Assume that $V_1 = \begin{pmatrix} x_1 \\ y_1 \end{pmatrix}$ and $V_2 = \begin{pmatrix} x_2 \\ y_2 \end{pmatrix}$ are eigenvectors of A, neither of which is a scalar multiple of the other. The system $X_0 = c_1V_1 + c_2V_2$ is equivalent to the algebraic system $\{c_1x_1 + c_2x_2 = x_0,\ c_1y_1 + c_2y_2 = y_0\}$, which has the solution $c_1 = -(x_0y_2 - x_2y_0)/(x_2y_1 - x_1y_2)$, $c_2 = (x_0y_1 - x_1y_0)/(x_2y_1 - x_1y_2)$, provided that $x_2y_1 - x_1y_2 \neq 0$. Now suppose that $x_2y_1 - x_1y_2 = 0$, or $x_2y_1 = x_1y_2$. If $y_1 \neq 0$, then $x_2 = \left(\frac{y_2}{y_1}\right)x_1$ and $y_2 = \left(\frac{y_2}{y_1}\right)y_1$. But then we have $V_2 = \left(\frac{y_2}{y_1}\right)\cdot V_1 = kV_1$, a contradiction of our assumption about V_1 and V_2. On the other hand, if $y_1 = 0$, we know that $x_1 \neq 0$ because V_1, as an eigenvector, can't be the zero vector. Then $x_2 = \left(\frac{x_2}{x_1}\right)x_1$ and $y_2 = \left(\frac{x_2}{x_1}\right)y_1$, so that $V_2 = \left(\frac{x_2}{x_1}\right)\cdot V_1 = kV_1$, another contradiction. The conclusion is that $x_2y_1 - x_1y_2 \neq 0$ and the algebraic system considered above does have a solution.

19. We have $\frac{dy}{dx} = \frac{\dot{y}}{\dot{x}} = \frac{y}{y} = 1$, provided that $y \neq 0$. Thus the slope of any trajectory not on the line determined by $V = \begin{pmatrix} 1 \\ 0 \end{pmatrix}$—that is, not on the x-axis—is a constant, 1. This implies that $y = x + k$, or $y(t) = x(t) + k$, for some constant k, so that all such trajectories form an infinite family of lines parallel to $y = x$.

5.4 Stability of Linear Systems: Equal Real Eigenvalues

1. (a) We can write the system as $\begin{pmatrix}\dot{x}\\ \dot{y}\end{pmatrix}=\begin{pmatrix}3 & 0\\ 0 & 3\end{pmatrix}\begin{pmatrix}x\\ y\end{pmatrix}$. The characteristic equation is $\lambda^2-6\lambda+9=(\lambda-3)^2=0$, so that the eigenvalues are $\lambda_1=3=\lambda_2$. The equation $\begin{pmatrix}\dot{x}\\ \dot{y}\end{pmatrix}=\begin{pmatrix}3 & 0\\ 0 & 3\end{pmatrix}\begin{pmatrix}x\\ y\end{pmatrix}=3\begin{pmatrix}x\\ y\end{pmatrix}=\begin{pmatrix}3x\\ 3y\end{pmatrix}$ is equivalent to the system $\{3x=3x,\ 3y=3y\}$, so that *any* nonzero vector is an eigenvector corresponding to the repeated eigenvalue $\lambda=3$. As in Example 5.4.1, we can take $V_1=\begin{pmatrix}1\\ 0\end{pmatrix}$ and $V_2=\begin{pmatrix}0\\ 1\end{pmatrix}$ to be two linearly independent eigenvectors.

(b) Here's a plot of the eigenvectors and some trajectories:

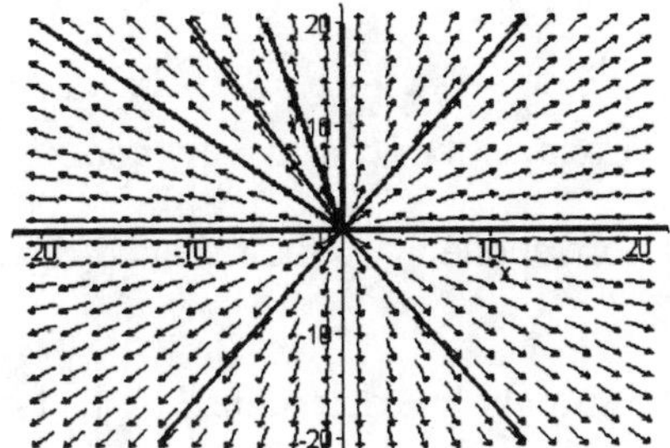

3. (a) We can write the system as $\begin{pmatrix}\dot{x}\\ \dot{y}\end{pmatrix}=\begin{pmatrix}2 & 1\\ -1 & 4\end{pmatrix}\begin{pmatrix}x\\ y\end{pmatrix}$. The characteristic equation is $\lambda^2-6\lambda+9=(\lambda-3)^2=0$, so that the eigenvalues are $\lambda_1=3=\lambda_2$. The equation $\begin{pmatrix}\dot{x}\\ \dot{y}\end{pmatrix}=\begin{pmatrix}2 & 1\\ -1 & 4\end{pmatrix}\begin{pmatrix}x\\ y\end{pmatrix}=3\begin{pmatrix}x\\ y\end{pmatrix}=\begin{pmatrix}3x\\ 3y\end{pmatrix}$ is equivalent to the system $\{2x+y=3x,\ -x+4y=3y\}$, or $\{-x+y=0,\ -x+y=0$. Therefore, we have only one equation, which can be written as $y=x$. Thus any nonzero vector of the form $\begin{pmatrix}x\\ x\end{pmatrix}=x\begin{pmatrix}1\\ 1\end{pmatrix}$ is an eigenvector corresponding to the repeated eigenvalue $\lambda=3$. This says that all eigenvectors lie on the straight line determined by $\begin{pmatrix}1\\ 1\end{pmatrix}$ and there is only one linearly independent eigenvector.

(b) Here's a plot of the eigenvector and some trajectories:

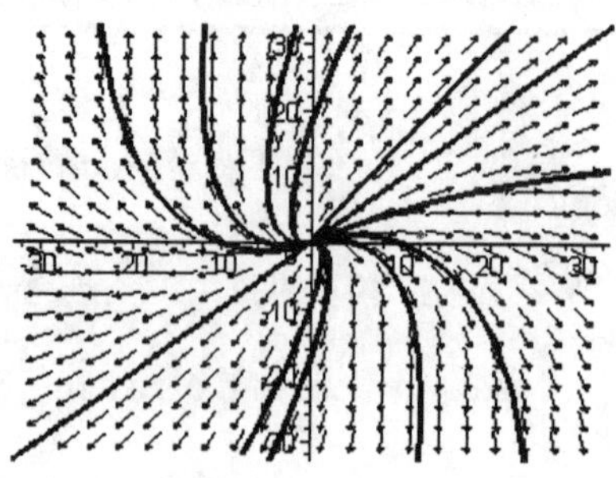

5. (a) We can write the system as $\begin{pmatrix}\dot{x}\\ \dot{y}\end{pmatrix}=\begin{pmatrix}-3 & 2\\ -2 & 1\end{pmatrix}\begin{pmatrix}x\\ y\end{pmatrix}$. The characteristic equation is $\lambda^2+2\lambda+1=(\lambda+1)^2=0$, so that the eigenvalues are $\lambda_1=-1=\lambda_2$. The equation $\begin{pmatrix}\dot{x}\\ \dot{y}\end{pmatrix}=\begin{pmatrix}-3 & 2\\ -2 & 1\end{pmatrix}\begin{pmatrix}x\\ y\end{pmatrix}=(-1)\begin{pmatrix}x\\ y\end{pmatrix}=\begin{pmatrix}-x\\ -y\end{pmatrix}$ is equivalent to the system $\{-3x+2y=-x,\ -2x+y=-y\}$, or $\{-2x+2y=0,\ -2x+2y=0\}$. Therefore, we have only one equation, which can be written as $y=x$. Thus any nonzero vector of the form $\begin{pmatrix}x\\ x\end{pmatrix}=x\begin{pmatrix}1\\ 1\end{pmatrix}$ is an eigenvector corresponding to the repeated eigenvalue $\lambda=-1$. This says that all eigenvectors lie on the straight line determined by $\begin{pmatrix}1\\ 1\end{pmatrix}$ and there is only one linearly independent eigenvector.

(b) Here's a plot of the eigenvector and some trajectories:

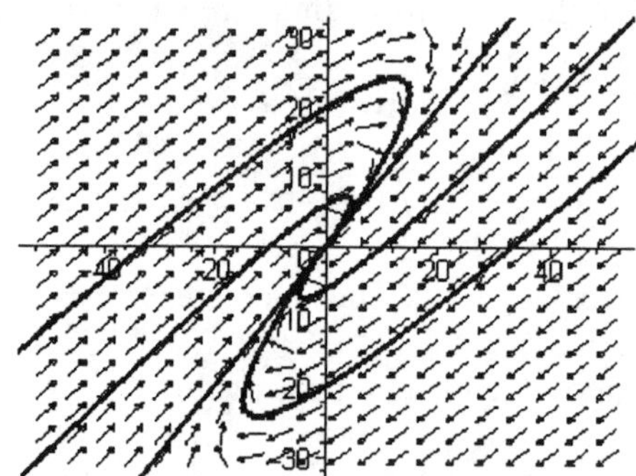

7. From the given eigenvalues, we construct the characteristic equation: $(\lambda-(-2))^2=(\lambda+2)^2$ $=\lambda^2+4\lambda+4=0$. This tells us that a matrix representing the system must have trace equal to -4 and determinant equal to 4. Two such matrices are $\begin{pmatrix}-2 & 0\\ 0 & -2\end{pmatrix}$ and $\begin{pmatrix}1 & 3\\ -3 & -5\end{pmatrix}$, yielding the systems $\{\dot{x}=-2x,\ \dot{y}=-2y\}$ and $\{\dot{x}=x+3y,\ \dot{y}=-3x-5y\}$.

9. (a) First of all, if $\lambda < 0$, then the origin is a *sink*—that is, all trajectories approach the origin as $t \to \infty$. Next note that $\frac{e^{-\lambda t}}{t}X(t)$, as an increasing scalar multiple of $X(t)$ for positive values of t, is parallel to $X(t)$. But $\frac{e^{-\lambda t}}{t}X(t) = \frac{e^{-\lambda t}}{t}\left[te^{\lambda t}\left(\frac{1}{t}(c_1V + W) + c_2V\right)\right]$ $= \frac{1}{t}(c_1V + W) + c_2V$, which tends to c_2V as $t \to \infty$, so that the slope of $X(t)$ approaches the slope of the line determined by V as $t \to \infty$. Specifically, any trajectory will be tangent to the vector V or its negative as it approaches the origin.

(b) Using the same reasoning and algebra as in part (a), we see that the slope of $X(t)$ approaches the slope of the line determined by V as $t \to -\infty$. In this case, however, $\frac{e^{-\lambda t}}{t}X(t)$ is a decreasing scalar multiple of $X(t)$ and the trajectories are moving backwards in time *away* from the origin. Here are two trajectories from Example 5.4.2:

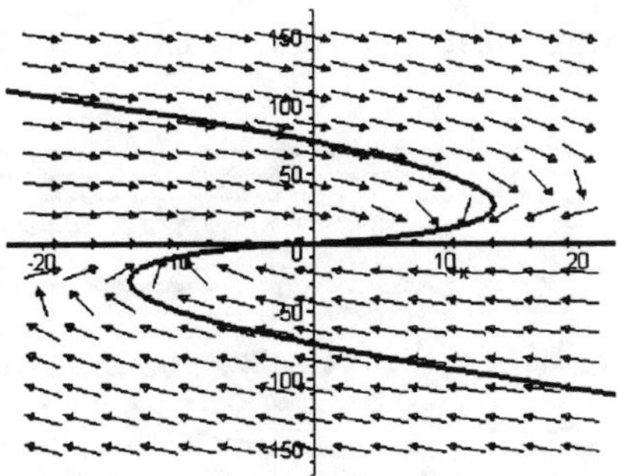

The only eigenvalue is $\lambda = -2$ and the sole eigenvector is $V = \begin{pmatrix} 1 \\ 0 \end{pmatrix}$, so that V and its negative determine the entire x-axis. Viewed in reverse time (as $t \to -\infty$), each trajectory is approaching a position parallel to either the positive x-axis or to the negative x-axis.

5.5 Stability of Linear Systems: Complex Eigenvalues

1. (a) The system can be written as $\begin{pmatrix} \dot{r} \\ \dot{s} \end{pmatrix} = \begin{pmatrix} -1 & -2 \\ 2 & -1 \end{pmatrix}\begin{pmatrix} r \\ s \end{pmatrix}$, and the characteristic equation is

$\lambda^2 + 2\lambda + 5 = 0$. Using the quadratic formula, we find that the eigenvalues are $\lambda_1 = -1 + 2i$ and $\lambda_2 = -1 - 2i$, a complex conjugate pair. For $\lambda = -1 + 2i$, we must have $\begin{pmatrix} -1 & -2 \\ 2 & -1 \end{pmatrix}\begin{pmatrix} r \\ s \end{pmatrix} = (-1+2i)\begin{pmatrix} r \\ s \end{pmatrix} = \begin{pmatrix} (-1+2i)r \\ (-1+2i)s \end{pmatrix}$, which is equivalent to the system

$$\begin{aligned} -r - 2s &= (-1+2i)r \\ 2r - s &= (-1+2i)s \end{aligned}, \quad \text{or} \quad \begin{aligned} -2ir - 2s &= 0 \\ 2r - 2is &= 0. \end{aligned}$$

Realizing that the second equation is just i times the first equation, we see that there is only one equation, which can be written as $s = -ir$. Therefore, any eigenvector corresponding to the complex eigenvalue $\lambda = -1 + 2i$ must have the form $V_1 = \begin{pmatrix} r \\ -ir \end{pmatrix} = r\begin{pmatrix} 1 \\ -i \end{pmatrix} = r\begin{pmatrix} i \\ 1 \end{pmatrix} = \begin{pmatrix} 0 \\ r \end{pmatrix} + i\begin{pmatrix} r \\ 0 \end{pmatrix} = U + iW$ for $r \neq 0$. Because eigenvectors corresponding to complex conjugate eigenvalues are conjugates of each other, we see that any eigenvector corresponding to $\lambda = -1 - 2i$ must have the form

$$V_2 = \overline{V_1} = r\overline{\begin{pmatrix} i \\ 1 \end{pmatrix}} = r\begin{pmatrix} -i \\ 1 \end{pmatrix} = \begin{pmatrix} 0 \\ r \end{pmatrix} - i\begin{pmatrix} r \\ 0 \end{pmatrix} = U - iW \text{ for } r \neq 0.$$

(b) Here's a plot of some trajectories, spirals swirling into the origin (a sink):

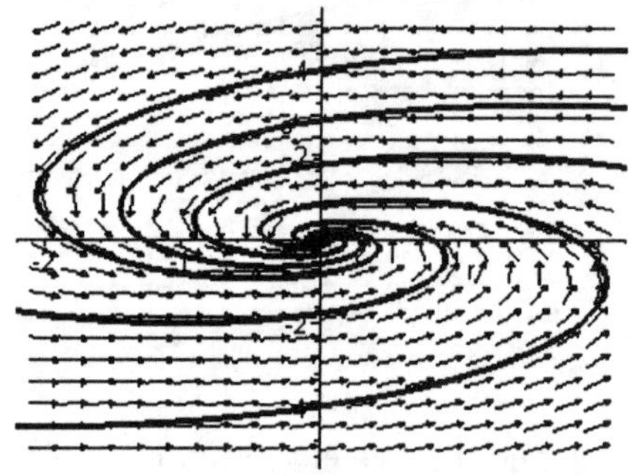

3. (a) The system is $\begin{pmatrix} \dot{x} \\ \dot{y} \end{pmatrix} = \begin{pmatrix} -0.5 & -1 \\ 1 & -0.5 \end{pmatrix}\begin{pmatrix} x \\ y \end{pmatrix}$ and the characteristic equation is

$\lambda^2 + \lambda + \frac{5}{4} = 0$. The eigenvalues are $\lambda_1 = -0.5 + i$ and $\lambda_2 = -0.5 - i$, with corresponding complex conjugate eigenvectors of the form $V_1 = x\begin{pmatrix} -1 \\ i \end{pmatrix} = x\begin{pmatrix} i \\ 1 \end{pmatrix}$ and $V_2 = x\begin{pmatrix} -i \\ 1 \end{pmatrix}$, $x \neq 0$.

(b) Some trajectories, spirals swirling toward the origin (a sink) follow:

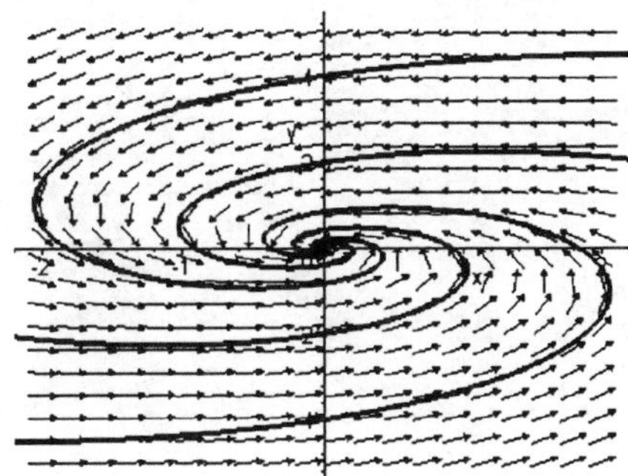

5. (a) The system is $\begin{pmatrix} \dot{x} \\ \dot{y} \end{pmatrix} = \begin{pmatrix} 2 & 1 \\ -3 & -1 \end{pmatrix} \begin{pmatrix} x \\ y \end{pmatrix}$ and the characteristic equation is $\lambda^2 - \lambda + 1 = 0$. The eigenvalues are $\lambda_1 = (1+\sqrt{3}\,i)/2$ and $\lambda_2 = (1-\sqrt{3}\,i)/2$, with corresponding complex conjugate eigenvectors of the form

$$V_1 = x\begin{pmatrix} 1 \\ (-3+\sqrt{3}\,i)/2 \end{pmatrix} \text{ and } V_2 = x\begin{pmatrix} 1 \\ (-3-\sqrt{3}\,i)/2 \end{pmatrix},\ x \neq 0.$$

(b) Here are some trajectories, spirals swirling away from the origin (a source:

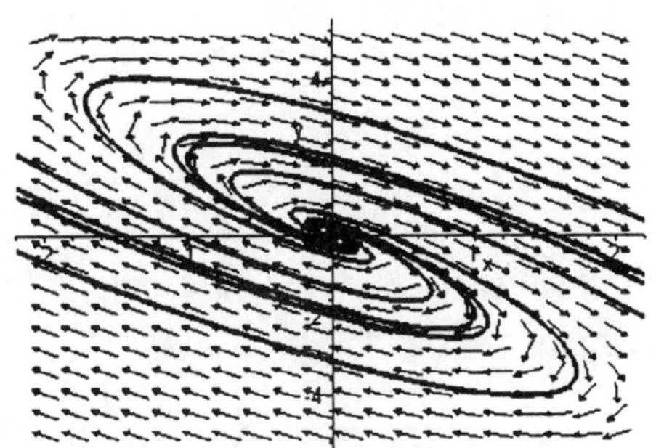

7. (a) The system is $\begin{pmatrix} \dot{x} \\ \dot{y} \end{pmatrix} = \begin{pmatrix} -7 & 1 \\ -2 & -5 \end{pmatrix} \begin{pmatrix} x \\ y \end{pmatrix}$ and the characteristic equation is $\lambda^2 + 12\lambda + 37 = 0$. The eigenvalues are $\lambda_1 = -6 + i$ and $\lambda_2 = -6 - i$, with corresponding complex conjugate eigenvectors of the form

$$V_1 = x\begin{pmatrix} (1-i)/2 \\ 1 \end{pmatrix} = x\begin{pmatrix} 1-i \\ 2 \end{pmatrix} \text{ and } V_2 = x\begin{pmatrix} 1+i \\ 2 \end{pmatrix},\ x \neq 0.$$

(b) Some trajectories, spirals swirling toward the origin (a sink) follow:

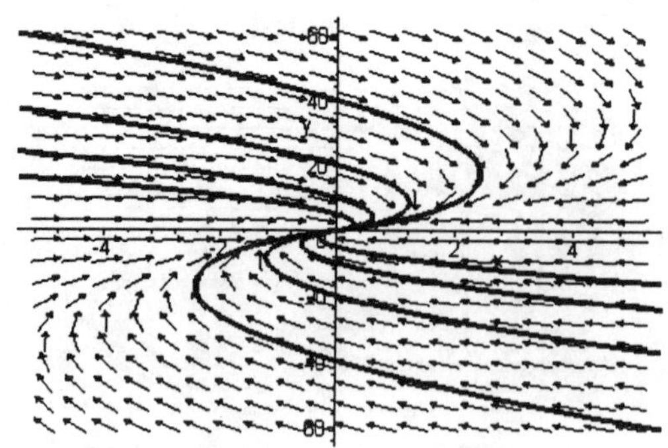

9. (a)

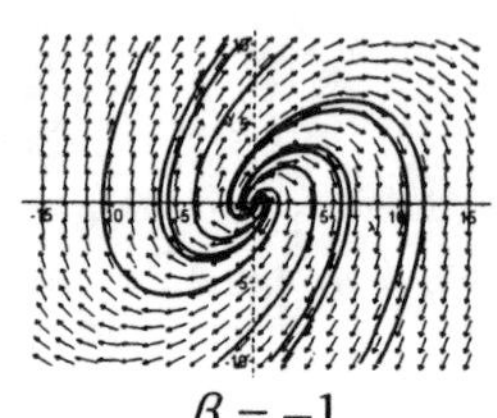
$\beta = -1$

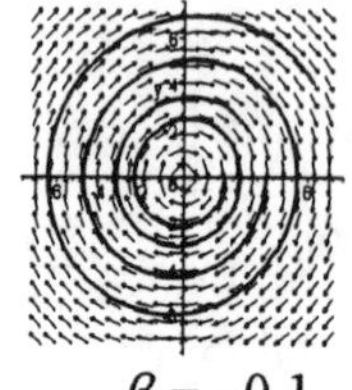
$\beta = -0.1$

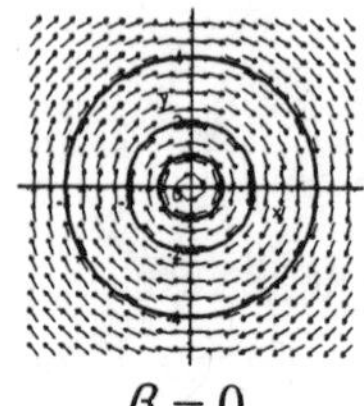
$\beta = 0$

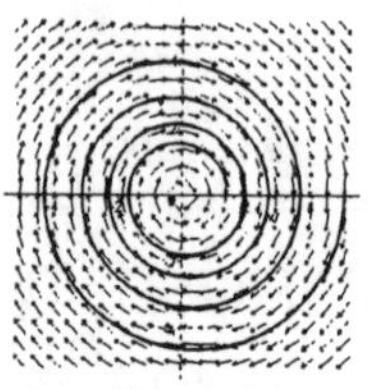
$\beta = 0.1$

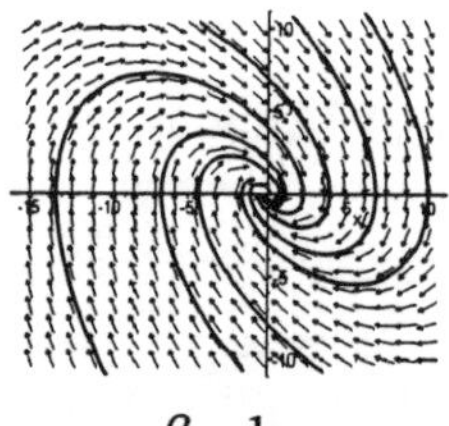
$\beta = 1$

As we go through the given values of β from left to right, we have a spiral source, a spiral source, a center, a spiral sink, and a spiral sink.

(b) Yes, $\beta = 0$ is the bifurcation point. As β passes through 0 from below, the origin changes from being a source to being a sink.

(c) The system can be written as $\begin{pmatrix} \dot{x} \\ \dot{y} \end{pmatrix} = \begin{pmatrix} 0 & 1 \\ -1 & -\beta \end{pmatrix} \begin{pmatrix} x \\ y \end{pmatrix}$, and its characteristic equation is $\lambda^2 + \beta\lambda + 1 = 0$. The eigenvalues are $\lambda_1 = \left(-\beta + \sqrt{\beta^2 - 4}\right)\Big/2$ and $\lambda_2 = \left(-\beta - \sqrt{\beta^2 - 4}\right)\Big/2$.

(d) When $\beta = -1$, $\lambda = \frac{1}{2} \pm \frac{\sqrt{3}}{2}i$, so that Table 5.1 indicates that we have a spiral source. When $\beta = -0.1$, $\lambda = 0.05 \pm 0.9987492980i$, so that Table 5.1 indicates that we have a spiral source. For $\beta = 0$, $\lambda = \pm\, i$, so that we have a center. When $\beta = 0.1$, $\lambda = -0.05 \pm 0.9987492980i$, so that the table gives us a spiral sink.

Finally, for $\beta=1$, $\lambda=-\frac{1}{2}\pm\frac{\sqrt{3}}{2}i$, and so we have a spiral sink.

11. First we note that if $X(t)=U+iW$, then $\dot{\overline{X}}=\overset{\bullet}{\overline{(U-iW)}}=\dot{U}-i\dot{W}=\overline{(\dot{U}+i\dot{W})}=\overline{\dot{X}}$. Now suppose that $X(t)$ is a complex-valued solution of $\dot{X}=AX$. Then

$$\dot{X}_1=\overset{\bullet}{\left(\frac{X-\overline{X}}{2i}\right)}=\frac{1}{2i}\left(\dot{X}-\dot{\overline{X}}\right)=\frac{1}{2i}\left(\dot{X}-\overline{\dot{X}}\right) \text{ and } AX_1=A\left(\frac{X-\overline{X}}{2i}\right)=\frac{1}{2i}A\left(X-\overline{X}\right)$$

$=\frac{1}{2i}\{A(X)-A(\overline{X})\}=\frac{1}{2i}\left\{\dot{X}-\overline{(AX)}\right\}=\frac{1}{2i}\left\{\dot{X}-\overline{\dot{X}}\right\}=\dot{X}_1$. Therefore, $X_1(t)$ is a solution of the system $\dot{X}=AX$.

5.6 Stability of Linear Systems: Nonhomogeneous Systems

1. As shown in the text, the general solution is $X(t)=\begin{pmatrix}x(t)\\y(t)\end{pmatrix}=\begin{pmatrix}c_1e^t-\frac{1}{2}(\sin t+\cos t)\\t-1+c_2e^{-t}\end{pmatrix}$. Then

$X(0)=\begin{pmatrix}x(0)\\y(0)\end{pmatrix}=\begin{pmatrix}0\\1\end{pmatrix}=\begin{pmatrix}c_1e^0-\frac{1}{2}(\sin 0+\cos 0)\\0-1+c_2e^{-0}\end{pmatrix}=\begin{pmatrix}c_1-\frac{1}{2}\\c_2-1\end{pmatrix}$, implying that $c_1=\frac{1}{2}$ and $c_2=2$.

Therefore the particular solution is $\{x(t)=\frac{1}{2}e^t-\frac{1}{2}(\sin t+\cos t),\ y(t)=t-1+2e^{-t}$.

3. We can write the nonhomogeneous system as $\begin{pmatrix}\dot{x}\\\dot{y}\end{pmatrix}=\begin{pmatrix}0&1\\1&0\end{pmatrix}\begin{pmatrix}x\\y\end{pmatrix}+\begin{pmatrix}2e^t\\t^2\end{pmatrix}=AX(t)+B(t)$. The eigenvalues of A are 1 and -1, with corresponding eigenvectors $V_1=\begin{pmatrix}1\\1\end{pmatrix}$ and $V_2=\begin{pmatrix}-1\\1\end{pmatrix}$.

Therefore, X_{GH}, the general solution of the homogeneous equation is

$\begin{pmatrix}c_1e^t-c_2e^{-t}\\c_1e^t+c_2e^{-t}\end{pmatrix}$. From Table 5.2 and the discussion preceding it, we choose a trial particular solution of the form $Ct^2+Dt+E+Fe^t+Gte^t$, where $C, D, E, F,$ and G are vectors of constants. Substituting the trial solution into the nonhomogeneous solution, we can solve for the constants to find that X_{PNH}, a particular solution of the nonhomogeneous equation is

$\begin{pmatrix}te^t-\frac{1}{2}e^t-t^2-2\\te^t-\frac{3}{2}e^t-2t\end{pmatrix}$. Then $X_{\mathrm{GNH}}=X_{\mathrm{GH}}+X_{\mathrm{PNH}}=\begin{pmatrix}c_1e^t-c_2e^{-t}+te^t-\frac{1}{2}e^t-t^2-2\\c_1e^t+c_2e^{-t}+te^t-\frac{3}{2}e^t-2t\end{pmatrix}$

$=\begin{pmatrix}\left(c_1-\frac{1}{2}\right)e^t-c_2e^{-t}+te^t-t^2-2\\\left(c_1-\frac{3}{2}\right)e^t+c_2e^{-t}+te^t-2t\end{pmatrix}$.

5. We can write the nonhomogeneous system as $\begin{pmatrix}\dot{x}\\ \dot{y}\end{pmatrix}=\begin{pmatrix}3 & 2\\ 1 & 2\end{pmatrix}\begin{pmatrix}x\\ y\end{pmatrix}+\begin{pmatrix}4e^{5t}\\ 0\end{pmatrix}=AX(t)+B(t)$.

The eigenvalues of A are 4 and 1, with corresponding eigenvectors $V_1=\begin{pmatrix}2\\ 1\end{pmatrix}$ and $V_2=\begin{pmatrix}-1\\ 1\end{pmatrix}$. Therefore, X_{GH}, the general solution of the homogeneous equation is $\begin{pmatrix}2c_1e^{4t}-c_2e^{t}\\ c_1e^{4t}+c_2e^{t}\end{pmatrix}$. From Table 5.2 and the discussion preceding it, we choose a trial particular solution of the form Ce^{5t}, where C is a vector of constants. Substituting the trial solution into the nonhomogeneous solution, we can solve for the constant to find that X_{PNH}, a particular solution of the nonhomogeneous equation is $\begin{pmatrix}3e^{5t}\\ e^{5t}\end{pmatrix}$. Then $X_{\text{GNH}}=X_{\text{GH}}+X_{\text{PNH}}=\begin{pmatrix}2c_1e^{4t}-c_2e^{t}+3e^{5t}\\ c_1e^{4t}+c_2e^{t}+e^{5t}\end{pmatrix}$.

7. We can write the nonhomogeneous system as $\begin{pmatrix}\dot{x}\\ \dot{y}\end{pmatrix}=\begin{pmatrix}4 & 1\\ -2 & 1\end{pmatrix}\begin{pmatrix}x\\ y\end{pmatrix}+\begin{pmatrix}-e^{2t}\\ 0\end{pmatrix}=AX(t)+B(t)$. The eigenvalues of A are 3 and 2, with corresponding eigenvectors $V_1=\begin{pmatrix}-1\\ 1\end{pmatrix}$ and $V_2=\begin{pmatrix}1\\ -2\end{pmatrix}$. Therefore, X_{GH}, the general solution of the homogeneous equation is $\begin{pmatrix}-c_1e^{3t}+c_2e^{2t}\\ c_1e^{3t}-2c_2e^{2t}\end{pmatrix}$. From Table 5.2 and the discussion preceding it, we choose a trial particular solution of the form $Ce^{2t}+Dte^{2t}$, where C and D are vectors of constants. Substituting the trial solution into the nonhomogeneous solution, we can solve for the constants to find that X_{PNH}, a particular solution of the nonhomogeneous equation is $\begin{pmatrix}te^{2t}\\ 2e^{2t}-2te^{2t}\end{pmatrix}$. Then $X_{\text{GNH}}=X_{\text{GH}}+X_{\text{PNH}}=\begin{pmatrix}-c_1e^{3t}+c_2e^{2t}+te^{2t}\\ c_1e^{3t}-2c_2e^{2t}+2e^{2t}-2te^{2t}\end{pmatrix}$

$$=\begin{pmatrix}-c_1e^{3t}+(t+c_2)e^{2t}\\ c_1e^{3t}-(2t-2+2c_2)e^{2t}\end{pmatrix}.$$

$$=\begin{pmatrix}-c_1e^{3t}+(t+c_2)e^{2t}\\ c_1e^{3t}-(2t-2+2c_2)e^{2t}\end{pmatrix}.$$

9. We can write the nonhomogeneous system as $\begin{pmatrix}\dot{x}\\ \dot{y}\end{pmatrix}=\begin{pmatrix}5 & -3\\ 1 & 1\end{pmatrix}\begin{pmatrix}x\\ y\end{pmatrix}+\begin{pmatrix}2e^{3t}\\ 5e^{-t}\end{pmatrix}=AX(t)+B(t)$. The eigenvalues of A are 4 and 2, with corresponding eigenvectors $V_1=\begin{pmatrix}3\\ 1\end{pmatrix}$ and $V_2=\begin{pmatrix}1\\ 1\end{pmatrix}$. Therefore, X_{GH}, the general solution of the homogeneous equation is

$\begin{pmatrix} 3c_1e^{4t} + c_2e^{2t} \\ c_1e^{4t} + c_2e^{2t} \end{pmatrix}$. From Table 5.2 and the discussion preceding it, we choose a trial particular solution of the form $Ce^{3t} + De^{-t}$, where C and D are vectors of constants. Substituting the trial solution into the nonhomogeneous solution, we can solve for the constants to find that X_{PNH}, a particular solution of the nonhomogeneous equation is $\begin{pmatrix} -4e^{3t} - e^{-t} \\ -2e^{3t} - 2e^{-t} \end{pmatrix}$. Then $X_{\text{GNH}} = X_{\text{GH}} + X_{\text{PNH}} = \begin{pmatrix} 3c_1e^{4t} + c_2e^{2t} - 4e^{3t} - e^{-t} \\ c_1e^{4t} + c_2e^{2t} - 2e^{3t} - 2e^{-t} \end{pmatrix}$.

11. We can write the nonhomogeneous system as $\begin{pmatrix} \dot{x} \\ \dot{y} \end{pmatrix} = \begin{pmatrix} 2 & -1 \\ -1 & 2 \end{pmatrix}\begin{pmatrix} x \\ y \end{pmatrix} + \begin{pmatrix} 0 \\ -5e^t \sin t \end{pmatrix} = AX(t) + B(t)$. The eigenvalues of A are 3 and 1, with corresponding eigenvectors $V_1 = \begin{pmatrix} -1 \\ 1 \end{pmatrix}$ and $V_2 = \begin{pmatrix} 1 \\ 1 \end{pmatrix}$. Therefore, X_{GH}, the general solution of the homogeneous equation is $\begin{pmatrix} -c_1e^{3t} + c_2e^t \\ c_1e^{3t} + c_2e^t \end{pmatrix}$. From Table 5.2 and the discussion preceding it, we choose a trial particular solution of the form $Ce^t \cos t + De^t \sin t$, where C and D are vectors of constants. Substituting the trial solution into the nonhomogeneous solution, we can solve for the constants to find that X_{PNH}, a particular solution of the nonhomogeneous equation is $\begin{pmatrix} 2e^t\cos t - e^t \sin t \\ 3e^t \cos t + e^t \sin t \end{pmatrix}$. Then

$$X_{\text{GNH}} = X_{\text{GH}} + X_{\text{PNH}} = \begin{pmatrix} -c_1e^{3t} + c_2e^t + 2e^t\cos t - e^t\sin t \\ c_1e^{3t} + c_2e^t + 3e^t\cos t + e^t\sin t \end{pmatrix}.$$

13. $\frac{dx}{dt} - x = \sin t \Rightarrow \mu(t) = e^{-t} \Rightarrow \frac{d}{dt}\left(e^{-t}x\right) = e^{-t}\sin t \Rightarrow e^{-t}x = \int e^{-t}\sin t\, dt$
$= -\frac{1}{2}e^{-t}\cos t - \frac{1}{2}e^{-t}\sin t + C \Rightarrow x(t) = -\frac{1}{2}\cos t - \frac{1}{2}\sin t + Ce^t$; and $\frac{dy}{dt} + y = t \Rightarrow$
$\mu(t) = e^t \Rightarrow \frac{d}{dt}\left(e^t y\right) = te^t \Rightarrow e^t y = \int te^t\, dt = te^t - e^t + K \Rightarrow y(t) = t - 1 + Ke^{-t}$.

15. (a) $AC + \begin{pmatrix} 0 \\ 2 \end{pmatrix} = \begin{pmatrix} 0 \\ 0 \end{pmatrix} \Rightarrow \begin{pmatrix} 0 & 1 \\ -2 & 3 \end{pmatrix}\begin{pmatrix} c_1 \\ c_2 \end{pmatrix} = \begin{pmatrix} 0 \\ -2 \end{pmatrix} \Rightarrow \begin{pmatrix} c_2 \\ -2c_1 + 3c_2 \end{pmatrix} = \begin{pmatrix} 0 \\ -2 \end{pmatrix} \Rightarrow c_2 = 0,\ c_1 = 1$
$\Rightarrow C = \begin{pmatrix} 1 \\ 0 \end{pmatrix}$.

(b) $2C = AD \Rightarrow \begin{pmatrix} 2 \\ 0 \end{pmatrix} = \begin{pmatrix} 0 & 1 \\ -2 & 3 \end{pmatrix}\begin{pmatrix} d_1 \\ d_2 \end{pmatrix} = \begin{pmatrix} d_2 \\ -2d_1 + 3d_2 \end{pmatrix} \Rightarrow d_2 = 2,\ d_1 = 3 \Rightarrow D = \begin{pmatrix} 3 \\ 2 \end{pmatrix}$.

(c) $D = AE \Rightarrow \begin{pmatrix} 3 \\ 2 \end{pmatrix} = \begin{pmatrix} 0 & 1 \\ -2 & 3 \end{pmatrix}\begin{pmatrix} e_1 \\ e_2 \end{pmatrix} = \begin{pmatrix} e_2 \\ -2e_1 + 3e_2 \end{pmatrix} \Rightarrow e_2 = 3,\ e_1 = 7/2 \Rightarrow E = \begin{pmatrix} 7/2 \\ 3 \end{pmatrix}.$

(d) $2G = AG \Rightarrow 2\begin{pmatrix} g_1 \\ g_2 \end{pmatrix} = \begin{pmatrix} 0 & 1 \\ -2 & 3 \end{pmatrix}\begin{pmatrix} g_1 \\ g_2 \end{pmatrix} = \begin{pmatrix} g_2 \\ -2g_1 + 3g_2 \end{pmatrix} \Rightarrow g_2 = 2g_1,$

where $g_1 \neq 0$ is arbitrary, $\Rightarrow G = \begin{pmatrix} g_1 \\ 2g_1 \end{pmatrix} = g_1\begin{pmatrix} 1 \\ 2 \end{pmatrix}$, where $g_1 \neq 0$ is arbitrary.

(e) $2F + G = AF + \begin{pmatrix} 0 \\ 3 \end{pmatrix} \Rightarrow \begin{pmatrix} 2f_1 \\ 2f_2 \end{pmatrix} + \begin{pmatrix} g_1 \\ 2g_1 \end{pmatrix} = \begin{pmatrix} 0 & 1 \\ -2 & 3 \end{pmatrix}\begin{pmatrix} f_1 \\ f_2 \end{pmatrix} = \begin{pmatrix} f_2 \\ -2f_1 + 3f_2 \end{pmatrix} \Rightarrow \begin{pmatrix} 2f_1 + g_1 \\ 2f_2 + 2g_1 \end{pmatrix}$

$= \begin{pmatrix} f_2 \\ -2f_1 + 3f_2 \end{pmatrix} \Rightarrow g_1 = f_2 - 2f_1$ and $2g_1 = -2f_1 + f_2 + 3 \Rightarrow g_1 = 3 \Rightarrow G = 3\begin{pmatrix} 1 \\ 2 \end{pmatrix} = \begin{pmatrix} 3 \\ 6 \end{pmatrix}.$

Then, letting $f_1 = 0$ for convenience, the equations above imply that $f_2 = 3$, so that $F = \begin{pmatrix} 0 \\ 3 \end{pmatrix}.$

17. (a) Finding the equilibrium point(s) is equivalent to solving the algebraic system $\{7y - 4x = 13,\ 2x - 5y = -11\}$, which has the solution $x = 2, y = 3$.

(b) The system can be written in the form $\begin{pmatrix} \dot{x} \\ \dot{y} \end{pmatrix} = \begin{pmatrix} -4 & 7 \\ 2 & -5 \end{pmatrix}\begin{pmatrix} x \\ y \end{pmatrix} + \begin{pmatrix} -13 \\ 11 \end{pmatrix} = AX + B.$

The characteristic equation of A is $\lambda^2 + 9\lambda + 6 = 0$, yielding the eigenvalues $\lambda_1 = \left(-9 + \sqrt{57}\right)/2$ and $\lambda_2 = \left(-9 - \sqrt{57}\right)/2$.

(c) The eigenvectors corresponding to λ_1 and λ_2 are $V_1 = \begin{pmatrix} \left(1 + \sqrt{57}\right)/4 \\ 1 \end{pmatrix} = \begin{pmatrix} 1 \\ \left(-1 + \sqrt{57}\right)/14 \end{pmatrix}$

and $V_2 = \begin{pmatrix} \left(1 - \sqrt{57}\right)/4 \\ 1 \end{pmatrix} = \begin{pmatrix} 1 \\ \left(-1 - \sqrt{57}\right)/14 \end{pmatrix}$, respectively.

(d) $X_{\text{GH}}(t) = c_1 e^{\left(-9 + \sqrt{57}\right)t/2}\begin{pmatrix} 1 \\ \frac{-1 + \sqrt{57}}{14} \end{pmatrix} + c_2 e^{\left(-9 - \sqrt{57}\right)t/2}\begin{pmatrix} 1 \\ \frac{-1 - \sqrt{57}}{14} \end{pmatrix}$

$$= \begin{pmatrix} c_1 e^{\left(-9 + \sqrt{57}\right)t/2} + c_2 e^{\left(-9 - \sqrt{57}\right)t/2} \\ \left(\frac{-1 + \sqrt{57}}{14}\right)c_1 e^{\left(-9 + \sqrt{57}\right)t/2} - \left(\frac{1 + \sqrt{57}}{14}\right)c_2 e^{\left(-9 - \sqrt{57}\right)t/2} \end{pmatrix}.$$

(e) Because the vector B in part (b) is a vector of constants, it is easy to see that any particular solution has to be a vector of constants. For example, $X_{\text{PNH}} = \begin{pmatrix} 2 \\ 3 \end{pmatrix}$ works.

(f) $X_{\text{GNH}} = \begin{pmatrix} c_1 e^{(\sqrt{57}-9)t/2} + c_2 e^{-(\sqrt{57}+9)t/2} \\ \left(\frac{-1+\sqrt{57}}{14}\right) c_1 e^{(\sqrt{57}-9)t/2} - \left(\frac{1+\sqrt{57}}{14}\right) c_2 e^{-(\sqrt{57}+9)t/2} \end{pmatrix} + \begin{pmatrix} 2 \\ 3 \end{pmatrix}$

$= \begin{pmatrix} c_1 e^{(\sqrt{57}-9)t/2} + c_2 e^{-(\sqrt{57}+9)t/2} + 2 \\ \left(\frac{-1+\sqrt{57}}{14}\right) c_1 e^{(\sqrt{57}-9)t/2} - \left(\frac{1+\sqrt{57}}{14}\right) c_2 e^{-(\sqrt{57}+9)t/2} + 3 \end{pmatrix}$. As $t \to \infty, X_{\text{GNH}} \to \begin{pmatrix} 2 \\ 3 \end{pmatrix}$

because all the exponential terms have negative exponents.

19. (a) The system can be written as $\begin{pmatrix} \dot{x} \\ \dot{y} \end{pmatrix} = \begin{pmatrix} -k_1 & 0 \\ k_1 & -k_2 \end{pmatrix} \begin{pmatrix} x \\ y \end{pmatrix} + \begin{pmatrix} I \\ 0 \end{pmatrix}$. The characteristic equation is $\lambda^2 + (k_1 + k_2)\lambda + k_1 k_2 = (\lambda + k_1)(\lambda + k_2) = 0$, so that the eigenvalues are $\lambda_1 = -k_1$ and $\lambda_2 = -k_2$. The associated eigenvectors are $V_1 = \begin{pmatrix} (k_2 - k_1)/k_1 \\ 1 \end{pmatrix}$ and $V_2 = \begin{pmatrix} 0 \\ 1 \end{pmatrix}$.

Therefore, $X_{\text{GH}} = c_1 e^{-k_1 t} \begin{pmatrix} (k_2 - k_1)/k_1 \\ 1 \end{pmatrix} + c_2 e^{-k_2 t} \begin{pmatrix} 0 \\ 1 \end{pmatrix} = \begin{pmatrix} \frac{k_2 - k_1}{k_1} c_1 e^{-k_1 t} \\ c_1 e^{-k_1 t} + c_2 e^{-k_2 t} \end{pmatrix}$. A trial particular solution of the nonhomogeneous equation should have the form of a vector of constants. Substituting a vector of the form $\begin{pmatrix} C_1 \\ C_2 \end{pmatrix}$ into the original nonhomogeneous equation, we find that $C_1 = I/k_1$ and $C_2 = I/k_2$, so that $X_{\text{PNH}} = \begin{pmatrix} I/k_1 \\ I/k_2 \end{pmatrix}$ and $X_{\text{GNH}} = \begin{pmatrix} \frac{k_2 - k_1}{k_1} c_1 e^{-k_1 t} + I/k_1 \\ c_1 e^{-k_1 t} + c_2 e^{-k_2 t} + I/k_2 \end{pmatrix}$. Using the initial conditions, we have $X_{\text{GNH}}(0) = \begin{pmatrix} x(0) \\ y(0) \end{pmatrix} = \begin{pmatrix} 0 \\ 0 \end{pmatrix} = \begin{pmatrix} \frac{k_2 - k_1}{k_1} c_1 + I/k_1 \\ c_1 + c_2 + I/k_2 \end{pmatrix}$, which yields $c_1 = I/(k_1 - k_2)$ and $c_2 = -k_1 I/(k_1 - k_2)$. Finally, we have

$$X_{\text{GNH}} = \begin{pmatrix} x(t) \\ y(t) \end{pmatrix} = \begin{pmatrix} -\frac{I}{k_1} e^{-k_1 t} + I/k_1 \\ \frac{I}{k_1 - k_2} e^{-k_1 t} - \frac{k_1 I}{k_2 (k_1 - k_2)} e^{-k_2 t} + I/k_2 \end{pmatrix}$$

$$= \begin{pmatrix} \frac{I}{k_1}\left(1 - e^{-k_1 t}\right) \\ \frac{I}{k_2}\left(1 + \frac{k_2}{k_1 - k_2} e^{-k_1 t} - \frac{k_1}{k_1 - k_2} e^{-k_2 t}\right) \end{pmatrix}.$$

(b) $\lim_{t \to \infty} x(t) = \lim_{t \to \infty} \frac{I}{k_1}\left(1 - e^{-k_1 t}\right) = \frac{I}{k_1}$ and $\lim_{t \to \infty} y(t) = \frac{I}{k_2}(1 + 0 - 0) = \frac{I}{k_2}$ because k_1 and k_2 in the exponents are positive constants.

(c) The graphs of $x(t)$ and $y(t)$ for the decongestant follow:

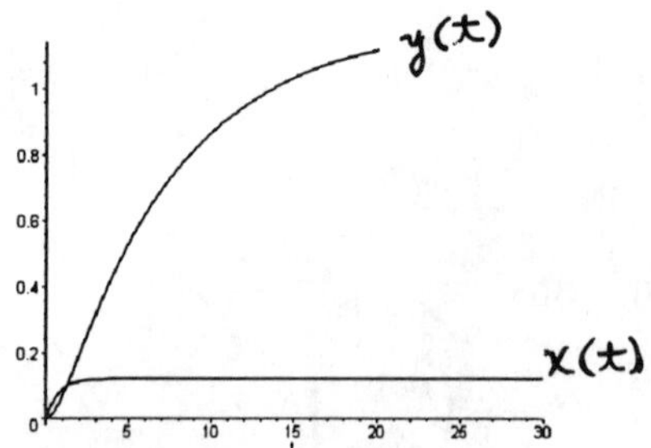

(d) The graphs of $x(t)$ and $y(t)$ for the antihistamine follow:

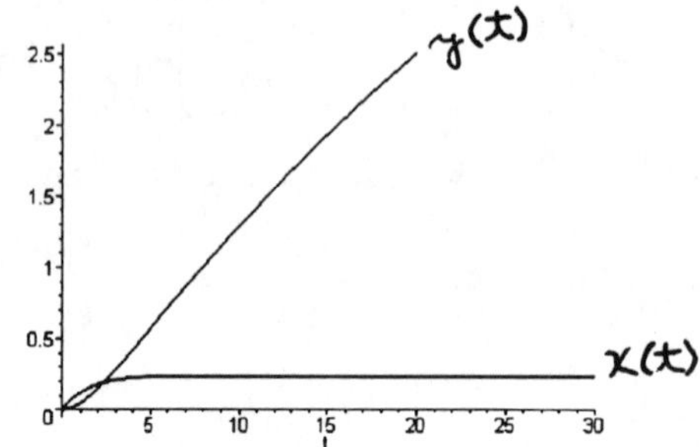

5.7 Generalizations: The $n \times n$ Case ($n \geq 3$)

1. (a) The system can be written as $\begin{pmatrix} \dot{x} \\ \dot{y} \\ \dot{z} \end{pmatrix} = \begin{pmatrix} 1 & -1 & 1 \\ 1 & 1 & -1 \\ 2 & -1 & 0 \end{pmatrix} \begin{pmatrix} x \\ y \\ z \end{pmatrix}$.

(b) The eigenvalue-eigenvector pairs are $\lambda_1 = -1, V_1 = \begin{pmatrix} 1 \\ -3 \\ -5 \end{pmatrix}; \lambda_2 = 1, V_2 = \begin{pmatrix} 1 \\ 1 \\ 1 \end{pmatrix}; \lambda_3 = 2, V_3 = \begin{pmatrix} 1 \\ 0 \\ 1 \end{pmatrix}$.

(c) $X(t) = \begin{pmatrix} c_1 e^{-t} + c_2 e^{t} + c_3 e^{2t} \\ -3c_1 e^{-t} + c_2 e^{t} \\ -5c_1 e^{-t} + c_2 e^{t} + c_3 e^{2t} \end{pmatrix}$.

3. (a) $\begin{pmatrix} \dot{x} \\ \dot{y} \\ \dot{z} \end{pmatrix} = \begin{pmatrix} 3 & -1 & 1 \\ 1 & 1 & 1 \\ 4 & -1 & 4 \end{pmatrix} \begin{pmatrix} x \\ y \\ z \end{pmatrix}$.

(b) The eigenvalue-eigenvector pairs are $\lambda_1 = 1, V_1 = \begin{pmatrix} 1 \\ 1 \\ -1 \end{pmatrix}; \lambda_2 = 2, V_2 = \begin{pmatrix} 1 \\ -2 \\ -3 \end{pmatrix}; \lambda_3 = 5, V_3 = \begin{pmatrix} 1 \\ 1 \\ 3 \end{pmatrix}$.

(c) $X(t)=\begin{pmatrix} c_1 e^{t} + c_2 e^{2t} + c_3 e^{5t} \\ c_1 e^{t} - 2c_2 e^{2t} + c_3 e^{5t} \\ -c_1 e^{t} - 3c_2 e^{2t} + 3c_3 e^{5t} \end{pmatrix}$.

5. The space trajectory through $(0, 1, 0)$ when $t = 0$ is

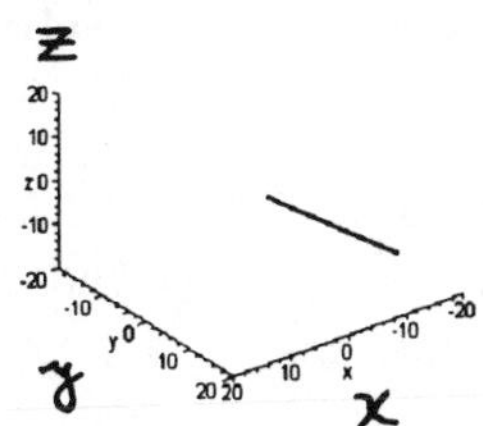

7. (a) The system can be written as $\begin{pmatrix} \dot{x} \\ \dot{y} \\ \dot{z} \end{pmatrix} = \begin{pmatrix} -1 & 1 & 1 \\ 1 & -1 & 1 \\ 1 & 1 & 1 \end{pmatrix}\begin{pmatrix} x \\ y \\ z \end{pmatrix}$. The eigenvalues are

$\lambda_1 = 2$, $\lambda_2 = -1$, and $\lambda_3 = -2$. The corresponding eigenvectors are $V_1 = \begin{pmatrix} 1 \\ 1 \\ 2 \end{pmatrix}$,

$V_2 = \begin{pmatrix} -1 \\ -1 \\ 1 \end{pmatrix}$, and $V_3 = \begin{pmatrix} -1 \\ 1 \\ 0 \end{pmatrix}$. Using the initial values, we find the solution

$$X(t) = \begin{pmatrix} x(t) \\ y(t) \\ z(t) \end{pmatrix} = \begin{pmatrix} \frac{2}{9}e^{-t} + \frac{2}{3}e^{-2t} + \frac{1}{9}e^{2t} \\ \frac{2}{9}e^{-t} - \frac{2}{3}e^{-2t} + \frac{1}{9}e^{2t} \\ -\frac{2}{9}e^{-t} + \frac{2}{9}e^{2t} \end{pmatrix}.$$

(b) Substituting in the solution given in part (a), we find that $x(0.5) \approx 0.6820688660$, $y(0.5) \approx 0.1915629444$, and $z(0.5) \approx 0.4692780374$.

(c) Using the *Improved Euler Method* with $h = 0.01$, we get the approximations $x(0.5) \approx 0.6820667761$, $y(0.5) \approx 0.1915276431$, and $z(0.5) \approx 0.4692372355$. With the *rkf45* method, we get $x(0.5) \approx 0.6820688625$, $y(0.5) \approx 0.1915629444$, and $z(0.5) \approx 0.4692780338$. Clearly the *rkf45* method produces more accurate approximations.

9. Denoting the volumes of tanks A, B, and C by $A(t)$, $B(t)$, and $C(t)$, respectively, the system is $\dot{X}(t) = \begin{pmatrix} -0.02 & 0.02 & 0 \\ 0.02 & -0.04 & 0.02 \\ 0 & 0.02 & -0.02 \end{pmatrix}\begin{pmatrix} A(t) \\ B(t) \\ C(t) \end{pmatrix}$. Note that this is a *closed system*—no fluid comes in from the outside, nor does fluid leave the system. The fact that a tank pumps fluid

back into itself does not change the volume of fluid in that tank. The characteristic equation is $\lambda^3 + 0.08\lambda^2 + 0.0012\lambda = 0$. The eigenvalue-eigenvector pairs are

$\lambda_1 = 0,\ V_1 = \begin{pmatrix} 1 \\ 1 \\ 1 \end{pmatrix}$; $\lambda_2 = -0.02,\ V_2 = \begin{pmatrix} -1 \\ 0 \\ 1 \end{pmatrix}$; $\lambda_3 = -0.06,\ V_3 = \begin{pmatrix} 1 \\ -2 \\ 1 \end{pmatrix}$. Using the initial conditions,

we find that the solution is $X(t) = \begin{pmatrix} 11000\,e^{-0.02t} + (11000/3)e^{-0.06t} + 25000/3 \\ -(22000/3)e^{-0.06t} + 25000/3 \\ -11000\,e^{-0.02t} + (11000/3)e^{-0.06t} + 25000/3 \end{pmatrix}$.

11. (a) Let $x_1(t)$ and $x_2(t)$ denote the amount of compound A in tanks I and II, respectively. Similarly, define $y_1(t)$, $y_2(t)$, $z_1(t)$, and $z_2(t)$ for compounds B and C in tanks I and II. Then the system is

$$\dot{X}(t) = \begin{pmatrix} -0.1 & 0.02 & 0 & 0 & 0 & 0 \\ 0.1 & -0.14 & 0 & 0 & 0 & 0 \\ 0 & 0 & -0.1 & 0.02 & 0 & 0 \\ 0 & 0 & 0.1 & -0.14 & 0 & 0 \\ 0 & 0 & 0 & 0 & -0.1 & 0.02 \\ 0 & 0 & 0 & 0 & 0.1 & -0.14 \end{pmatrix} \begin{pmatrix} x_1(t) \\ x_2(t) \\ y_1(t) \\ y_2(t) \\ z_1(t) \\ z_2(t) \end{pmatrix} + \begin{pmatrix} 4 \\ 2 \\ 0 \\ 0 \\ 0 \\ 0 \end{pmatrix},$$

with $x_1(0) = 0$, $x_2(0) = 0$, $y_1(0) = 50$, $y_2(0) = 0$, $z_1(0) = 0$, and $z_2(0) = 50$.

(b) $x_1(t) = 50 - 4.59e^{-0.169t} - 45.41e^{-0.071t}$,
$x_2(t) = 50 + 15.82e^{-0.169t} - 65.82e^{-0.071t}$,
$y_1(t) = 14.79e^{-0.169t} + 35.21e^{-0.071t}$, $y_2(t) = -51.03e^{-0.169t} + 51.03e^{-0.071t}$,
$z_1(t) = -10.21e^{-0.169t} + 10.21e^{-0.071t}$, $z_2(t) = 35.21e^{-0.169t} + 14.79e^{-0.071t}$.

(c) Here are $x_1(t)$, $y_1(t)$, and $z_1(t)$ on the same set of axes:

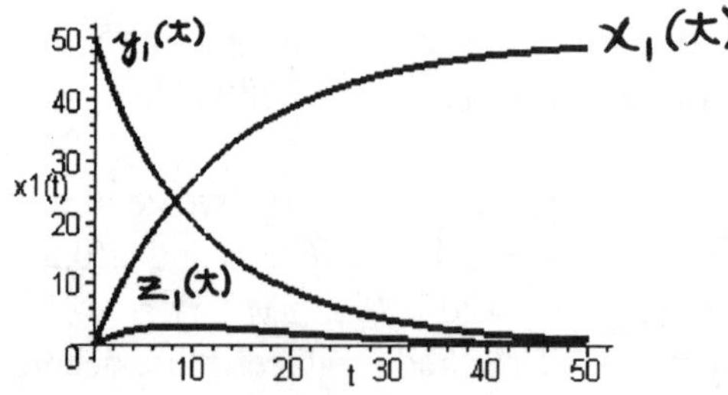

(d) Here are $x_2(t)$, $y_2(t)$, and $z_2(t)$ on the same set of axes:

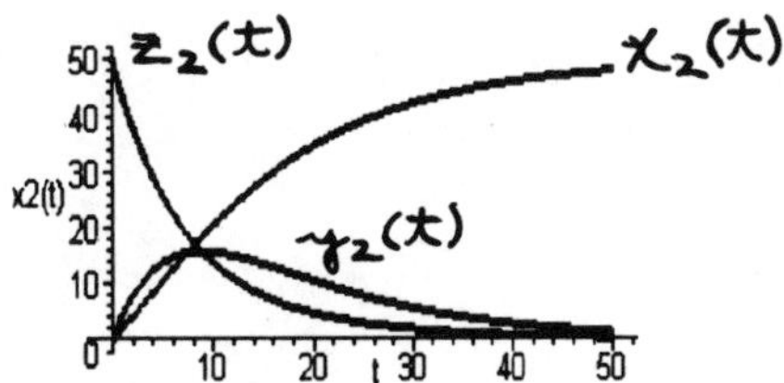

13. Using the initial conditions in Example 5.7.7, we have $X(0)=\begin{pmatrix} x_1(0) \\ x_2(0) \\ x_3(0) \\ x_4(0) \end{pmatrix}=\begin{pmatrix} 0 \\ 0 \\ 0 \\ 1 \end{pmatrix}=\begin{pmatrix} 3c_1+2c_3 \\ 3\sqrt{15}\,c_2+2\sqrt{2}\,c_4 \\ -2c_1+3c_3 \\ -2\sqrt{15}\,c_2+3\sqrt{2}\,c_4 \end{pmatrix}$. Working first with the first and third equations, then with the second and fourth, we find that $c_3=0=c_1$, $c_4=3\sqrt{2}/26$, and $c_2=-2\sqrt{15}/195$.

15. (a) The system can be written as $\begin{pmatrix} \dot{x} \\ \dot{y} \\ \dot{z} \end{pmatrix}=\begin{pmatrix} 0 & 0 & 0 \\ 3 & 0 & 0 \\ 1 & 4 & 0 \end{pmatrix}\begin{pmatrix} x \\ y \\ z \end{pmatrix}+\begin{pmatrix} 2t \\ 2t \\ t \end{pmatrix}$. First we find the general solution of the homogeneous equation. Technology gives us the eigenvalues $\lambda_1=\lambda_2=\lambda_3=0$, with only one eigenvector, $V=\begin{pmatrix} 0 \\ 0 \\ 1 \end{pmatrix}$. As in Exercise 14, we can find two additional linearly independent vectors W and U, which in this case are $W=\begin{pmatrix} 0 \\ 1/4 \\ 0 \end{pmatrix}$ and $U=\begin{pmatrix} 1/12 \\ -1/48 \\ 0 \end{pmatrix}$. Putting the components together as in Exercise 14, we conclude that $X_{\text{GH}}=\begin{pmatrix} c_3/12 \\ c_3\left(\frac{t}{4}-\frac{1}{48}\right)+\frac{c_2}{4} \\ c_3\frac{t^2}{2}+c_2 t+c_1 \end{pmatrix}$. Using the technique of undetermined coefficients, we find that $X_{\text{PNH}}=\begin{pmatrix} t^2 \\ t^3+t^2 \\ t^4+\frac{5}{3}t^3+\frac{1}{2}t^2 \end{pmatrix}$. Then $X_{\text{GNH}}=X_{\text{GH}}+X_{\text{PNH}}=$

$$\begin{pmatrix} t^2+\frac{c_3}{12} \\ t^3+t^2+\frac{c_3}{4}t+\frac{c_2}{4}-\frac{c_3}{48} \\ t^4+\frac{5}{3}t^3+\frac{1}{2}(c_3+1)t^2+c_2 t+c_1 \end{pmatrix}.$$

(b) We solve each equation, substitute the solution in the following equation, and then integrate: $\dot{x}=2t \Rightarrow x=t^2+C_1$; $\dot{y}=3x+2t=3\left(t^2+C_1\right)+2t \Rightarrow y=t^3+t^2+3C_1 t+C_2$; $\dot{z}=x+4y+t=\left(t^2+C_1\right)+4\left(t^3+t^2+3C_1 t+C_2\right)+t \Rightarrow$

$$z=t^4+\tfrac{5}{3}t^3+(12C_1+1)\frac{t^2}{2}+(4C_2+C_1)t+C_3.$$

Chapter 6

The Laplace Transform

6.1 The Laplace Transform of Some Important Functions

1. Integrating by parts, we get $\int_0^{\infty} t^2 e^{-st}\,dt = \lim_{b\to\infty}\left\{\left(\frac{-t^2}{s}e^{-st}\right]_0^b - \int_0^b\left(\frac{-1}{s}e^{-st}\right)2t\,dt\right\}$

$\frac{2}{s}\left\{\lim_{b\to\infty}\int_0^b te^{-st}\,dt\right\} = \frac{2}{s}\mathcal{L}[t] = \frac{2}{s}\cdot\frac{1}{s^2} = \frac{2}{s^3},\ s>0.$

3. $\mathcal{L}[\cos(at)] = \int_0^{\infty}\cos(at)e^{-st}\,dt = \lim_{b\to\infty}\left\{\left(\frac{-\cos(at)}{s}e^{-st}\right]_0^b - \int_0^b\left(\frac{-1}{s}e^{-st}\right)(-a\sin(at))dt\right\} =$

$= \frac{1}{s} - \frac{a}{s}\ \mathcal{L}[\sin(at)] = \frac{1}{s} - \frac{a}{s}\cdot\left(\frac{a}{s^2+a^2}\right) = \frac{s^2}{s\left(s^2+a^2\right)} = \frac{s}{s^2+a^2}$ for $s>0$.

5. $\int_0^{\infty}\left(\frac{e^{at}-e^{bt}}{a-b}\right)e^{-st}\,dt = \frac{1}{a-b}\left\{\int_0^{\infty}e^{at}e^{-st}dt - \int_0^{\infty}e^{bt}e^{-st}dt\right\} = \frac{1}{a-b}\left\{\mathcal{L}\left[e^{at}\right] - \mathcal{L}\left[e^{bt}\right]\right\}$

$= \frac{1}{a-b}\left\{\frac{1}{s-a} - \frac{1}{s-b}\right\} = \frac{1}{(s-a)(s-b)}$ for $s>a$ and $s>b$.

7. $\mathcal{L}[A(t)] = \int_0^1 1\cdot e^{-st}\,dt + \int_1^2(2-t)e^{-st}\,dt + \int_2^{\infty}0\cdot e^{-st}\,dt = \int_0^1 e^{-st}\,dt\ +\ \int_1^2 2e^{-st}\,dt\ -\ \int_1^2 te^{-st}\,dt =$

$\left(-\frac{e^{-s}}{s}+\frac{1}{s}\right) + 2\left(-\frac{e^{-2s}}{s}+\frac{e^{-s}}{s}\right) - \left\{-\frac{2e^{-2s}}{s} + \frac{e^{-s}}{s} + \frac{1}{s}\left(-\frac{e^{-2s}}{s} + \frac{e^{-s}}{s}\right)\right\}$

$= \frac{1}{s} + \frac{e^{-2s}}{s^2} - \frac{e^{-s}}{s^2} = \frac{s + e^{-2s} - e^{-s}}{s^2}.$

9. $\mathcal{L}[y'-y] = \mathcal{L}[0] \Rightarrow \mathcal{L}[y'] - \mathcal{L}[y] = 0 \Rightarrow s\mathcal{L}[y] - y(0) - \mathcal{L}[y] = (s-1)\mathcal{L}[y] - 1 = 0$

$\Rightarrow \mathcal{L}[y] = \frac{1}{s-1}.$

11. $\mathcal{L}[y''+4y'+4y] = \mathcal{L}[0] \Rightarrow \mathcal{L}[y''] + 4\mathcal{L}[y'] + 4\mathcal{L}[y] = 0 \Rightarrow s^2\mathcal{L}[y] - s\,y(0) - y'(0)$

$+4(s\mathcal{L}[y] - y(0)) + 4\mathcal{L}[y] = 0 \Rightarrow \left(s^2+4s+4\right)\mathcal{L}[y] - s - 5 = 0 \Rightarrow \mathcal{L}[y] = \frac{s+5}{(s+2)^2}.$

13. $\mathcal{L}[y''' - 2y'' + y'] = \mathcal{L}[2e^x + 2x] \Rightarrow \mathcal{L}[y'''] - 2\mathcal{L}[y''] + \mathcal{L}[y'] = \dfrac{2}{s-1} + \dfrac{2}{s^2} \Rightarrow$
$(s^3\mathcal{L}[y] - s^2 y(0) - s\,y'(0) - y''(0)) - 2(s^2\mathcal{L}[y] - s\,y(0) - y'(0)) + (s\mathcal{L}[y] - y(0)) =$
$\dfrac{2}{s-1} + \dfrac{2}{s^2} \Rightarrow (s^3 - 2s^2 + s)\mathcal{L}[y] = \dfrac{2}{s-1} + \dfrac{2}{s^2} \Rightarrow \mathcal{L}[y] = \dfrac{2(s^2+s-1)}{s^3(s-1)^3}.$

15. $\mathcal{L}[t^{-1/2}] = \int_0^\infty t^{-1/2} e^{-st}\,dt = \lim_{b\to\infty} \int_0^b t^{-1/2} e^{-st}\,dt$. Make the substitution $t = \dfrac{u^2}{s}$, so that $dt = \dfrac{2u}{s}du$. Then $t = 0 \Rightarrow u = 0$ and $t = b \Rightarrow u = \sqrt{sb}$. Our integral becomes
$\lim_{b\to\infty} \int_0^{\sqrt{sb}} \left(\dfrac{u^2}{s}\right)^{-1/2} e^{-st} \dfrac{2u}{s}\,du = \dfrac{2\sqrt{s}}{s} \cdot \lim_{b\to\infty} \int_0^{\sqrt{sb}} e^{-u^2}\,du = \dfrac{2\sqrt{s}}{s}\int_0^\infty e^{-u^2}\,du = \dfrac{2\sqrt{s}}{s}\cdot\dfrac{\sqrt{\pi}}{2}$
$= \dfrac{\sqrt{s}}{s}\cdot\sqrt{\pi} = \dfrac{1}{\sqrt{s}}\cdot\sqrt{\pi} = \sqrt{\dfrac{\pi}{s}}.$

17. First, we have $\mathcal{L}[e^{at}f(t)] = \int_0^\infty e^{at} f(t)\, e^{-st}\,dt = \int_0^\infty f(t)\, e^{(a-s)t}\,dt = \int_0^\infty f(t)\, e^{-(s-a)t}\,dt$.
Then $F(s) = \mathcal{L}[f(t)] = \int_0^\infty f(t)\, e^{-st}\,dt$ and, substituting $s-a$ for s, we have
$F(s-a) = \int_0^\infty f(t)\, e^{-(s-a)t}\,dt = \mathcal{L}[e^{at}f(t)]$ for $s > a$.

6.2 The Inverse Transform and the Convolution

1. If $F(s) = \mathcal{L}[f(t)]$ and $G(s) = \mathcal{L}[g(t)]$, then $\mathcal{L}^{-1}[F(s)] = f(t)$ and $\mathcal{L}^{-1}[G(s)] = g(t)$.
Furthermore, $\mathcal{L}^{-1}[c_1F(s) + c_2G(s)] = \mathcal{L}^{-1}[c_1\mathcal{L}[f(t)] + c_2\mathcal{L}[g(t)]]$
$= \mathcal{L}^{-1}[\mathcal{L}[c_1 f(t)] + \mathcal{L}[c_2 g(t)]]$ = [by (6.1.3)] $\mathcal{L}^{-1}[\mathcal{L}[c_1 f(t) + c_2 g(t)]]$
$= c_1 f(t) + c_2 g(t) = c_1\mathcal{L}^{-1}[F(s)] + c_2\mathcal{L}^{-1}[G(s)]$.

3. Using the technique of partial fractions, and entries 5, 6, and 9 of Table 6.1, we can write
$F(s) = \dfrac{1}{s(s^2+2s+2)} = \dfrac{\frac{1}{2}}{s} - \dfrac{\frac{1}{2}s+1}{(s+1)^2+1} = \dfrac{\frac{1}{2}}{s} - \dfrac{s+1}{(s+1)^2+1} + \dfrac{\frac{1}{2}s}{(s+1)^2+1}$, so that
$f(t) = \mathcal{L}^{-1}[F(s)] = \frac{1}{2}\mathcal{L}^{-1}\left[\dfrac{1}{s}\right] - \mathcal{L}^{-1}\left[\dfrac{s-(-1)}{(s-(-1))^2+1}\right] + \frac{1}{2}\mathcal{L}^{-1}[s\cdot\mathcal{L}[e^{-t}\sin t]]$
$= \dfrac{1}{2} - e^{-t}\cos t + \dfrac{1}{2}\left\{(e^{-t}\sin t)' + e^{-0}\sin(0)\right\} = \dfrac{1}{2} - e^{-t}\cos t + \dfrac{1}{2}\left\{e^{-t}\cos t - e^{-t}\sin t\right.$
$= \dfrac{1}{2} - \dfrac{1}{2}e^{-t}(\cos t + \sin t).$

5. (a) We have $F(s)=\int_0^\infty f(t)e^{-st}$. Then, by *Leibniz's rule for differentiating under an integral sign,* $\frac{d}{ds}F(s)=\int_0^\infty \frac{\partial}{\partial s}\{f(t)e^{-st}\}dt=\int_0^\infty f(t)\frac{\partial}{\partial s}\{e^{-st}\}dt=-\int_0^\infty t\,f(t)e^{-st}\,dt=-\mathcal{L}\,[t\,f(t)]$, so that $(-1)^1F'(s)=\mathcal{L}\left[t^1\,f(t)\right]$. The general formula follows by mathematical induction.

(b) $F(s)\,=\,\ln\left(2+\frac{3}{s}\right)\Rightarrow F'(s)=\frac{1}{2+\frac{3}{s}}\cdot\left(-3s^{-2}\right)=-\frac{3}{s(2s+3)}$. Then $\mathcal{L}^{-1}[F'(s)]$

$=\,\mathcal{L}^{-1}\left[\frac{2}{2s+3}-\frac{1}{s}\right]=2\mathcal{L}^{-1}\left[\frac{1}{2\left(s+\frac{3}{2}\right)}\right]-\mathcal{L}^{-1}\left[\frac{1}{s}\right]=\mathcal{L}^{-1}\left[\frac{1}{s-\left(-\frac{3}{2}\right)}\right]-\mathcal{L}^{-1}\left[\frac{1}{s}\right]$

$=\,e^{-3t/2}\,-\,1$. By part (a), this last expression equals $-t\,f(t)$, so

$f(t)\,=\,-\frac{1}{t}\left(e^{-3t/2}\,-\,1\right)\,=\,\frac{1-e^{-3t/2}}{t}$.

7. (a) $(f*g)(t)=\,\int_0^t f(r)\cdot g(t-r)\,dr\,=\,[\text{letting } u=t-r]\,-\int_t^0 f(t-u)g(u)\,du$

$=\,\int_0^t f(t-u)g(u)\,du\,=\,\int_0^t g(u)f(t-u)\,du\,=\,(g*f)(t)$.

(b) By the Convolution Theorem, $\mathcal{L}[f*(g*h)]=\mathcal{L}[f]\cdot\mathcal{L}[g*h]=\mathcal{L}[f]\cdot\mathcal{L}[g]\cdot\mathcal{L}[h]$ $=\{\mathcal{L}[f]\cdot\mathcal{L}[g]\}\cdot\mathcal{L}[h]=\mathcal{L}[f*g]\cdot\mathcal{L}[h]=\mathcal{L}[(f*g)*h]$. Applying the inverse Laplace transform, we see that $f*(g*h)=(f*g)*h$.

(c) $[f*(g+h)](t)=\int_0^t f(r)[g(t-r)+h(t-r)]\,dr=\int_0^t f(r)g(t-r)+f(r)h(t-r)\;dr$

$=\int_0^t f(r)g(t-r)\;dr+\int_0^t f(r)h(t-r)\;dr=(f*g)(t)+(f*h)(t)$.

(d) $(f*0)(t)=\int_0^t f(r)\cdot 0\;dr=0$, but $(f*1)(t)=\int_0^t f(r)\,dr\neq f(t)$ in general. For example, let $f(t)=t$. Then $(f*1)(t)=\int_0^t r\;dr=\frac{t^2}{2}\neq t$. (Also see problem 19 in Exercises 2.1.)

Furthermore $(f*f)(t)=\int_0^t f(r)\cdot f(t-r)\;dr\neq f^2(t)$ in general, as the example $f(t)=t$ shows: $(t*t)=\int_0^t r\cdot(t-r)\,dr=\frac{t^3}{6}\neq t^2$. (Also see problem 6(d) in this set of exercises.)

Finally, we note that $(1*1)=\int_0^t 1\cdot 1\;dr=t$.

9. We want $\mathcal{L}[f(t)] = \mathcal{L}\left[\int_0^t \cos(t-r)\sin r\, dr\right]$. Recognizing that the integral is $(\sin * \cos)(t)$, the convolution of $g(t) = \sin t$ and $h(t) = \cos t$, we use the Convolution Theorem to find that $\mathcal{L}\left[\int_0^t \cos(t-r)\sin r\, dr\right] = \mathcal{L}[(g*h)(t)] = \mathcal{L}[g(t)]\,\mathcal{L}[h(t)]$

= [entries 3 and 4 in Table 6.1] $\frac{1}{s^2+1}\cdot\frac{s}{s^2+1} = \frac{s}{\left(s^2+1\right)^2}$.

11. $\mathcal{L}[y''+3y'+2y] = \mathcal{L}[4t^2] \Rightarrow \mathcal{L}[y''] + 3\mathcal{L}[y'] + 2\mathcal{L}[y] = 4\mathcal{L}[t^2]$

$\Rightarrow \{s^2\mathcal{L}[y] - s\,y(0) - y'(0)\} + 3\{s\mathcal{L}[y] - y(0)\} + 2\mathcal{L}[y] = \frac{8}{s^3}$

$\Rightarrow (s^2+3s+2)\mathcal{L}[y] = \frac{8}{s^3} \Rightarrow \mathcal{L}[y] = \frac{8}{s^3(s^2+3s+2)} = \frac{8}{s^3(s+2)(s+1)}$

$= \frac{4}{s^3} - \frac{6}{s^2} + \frac{7}{s} + \frac{1}{s+2} - \frac{8}{s+1} \Rightarrow y(t) = \mathcal{L}^{-1}\left[\frac{4}{s^3} - \frac{6}{s^2} + \frac{7}{s} + \frac{1}{s+2} - \frac{8}{s+1}\right]$

$= 4\mathcal{L}^{-1}\left[\frac{1}{s^3}\right] - 6\mathcal{L}^{-1}\left[\frac{1}{s^2}\right] + 7\mathcal{L}^{-1}\left[\frac{1}{s}\right] + \mathcal{L}^{-1}\left[\frac{1}{s+2}\right] - 8\mathcal{L}^{-1}\left[\frac{1}{s+1}\right]$

$= 2t^2 - 6t + 7 + e^{-2t} - 8e^{-t}$.

13. $\mathcal{L}[y'''-2y''+y'] = \mathcal{L}[2e^x+2x] \Rightarrow \mathcal{L}[y'''] - 2\mathcal{L}[y''] + \mathcal{L}[y'] = \mathcal{L}[2e^x+2x]$

$\Rightarrow \{s^3\mathcal{L}[y] - s^2y(0) - s\,y'(0) - y''(0)\} - 2\{s^2\mathcal{L}[y] - s\,y(0) - y'(0)\} + \{s\mathcal{L}[y] - y(0)\}$

$= \frac{2}{s-1} + \frac{2}{s^2} \Rightarrow (s^3 - 2s^2 + s)\mathcal{L}[y] = \frac{2s^2+2s-2}{s^2(s-1)^3} \Rightarrow \mathcal{L}[y] = \frac{2(s^2+s-1)}{s^3(s-1)^3}$

$= \frac{2}{s^3} + \frac{4}{s^2} + \frac{4}{s} + \frac{2}{(s-1)^3} - \frac{4}{s-1} \Rightarrow y(x) = \mathcal{L}^{-1}\left[\frac{2}{s^3}\right] + 4\mathcal{L}^{-1}\left[\frac{1}{s^2}\right] + 4\mathcal{L}^{-1}\left[\frac{1}{s}\right] + \mathcal{L}^{-1}\left[\frac{2}{(s-1)^3}\right]$

$-4\mathcal{L}^{-1}\left[\frac{1}{(s-1)}\right] = x^2 + 4x + 4 + x^2e^x - 4e^x$.

15. Note that if $f(t) = \{0 \text{ for } 0 \le t < \pi \text{ and } -\sin t \text{ for } t \ge \pi\}$, then

$\mathcal{L}[f(t)] = \left\{0 \text{ for } 0 \le t < \pi \text{ and } \int_\pi^\infty (-\sin t)e^{-st}\,dt = \frac{e^{-\pi s}}{s^2+1} \text{ for } t \ge \pi\right\}$. Now

$\mathcal{L}[Q''+2Q'+2Q] = \{0 \text{ for } 0 \le t < \pi;\ \mathcal{L}[-\sin t] \text{ for } t \ge \pi\} \Rightarrow \mathcal{L}[Q''] + 2\mathcal{L}[Q'] + 2\mathcal{L}[Q]$

$= \{0 \text{ for } 0 \le t < \pi;\ e^{-\pi s}/(s^2+1) \text{ for } t \ge \pi\} \Rightarrow \{s^2\mathcal{L}[Q] - sQ(0) - Q'(0)\} + 2\{s\mathcal{L}[Q] - Q(0)\}$
$+ 2\mathcal{L}[Q] = \{0 \text{ for } 0 \le t < \pi;\ e^{-\pi s}/(s^2+1) \text{ for } t \ge \pi\} \Rightarrow (s^2+2s+2)\mathcal{L}[Q] =$
$\{1 \text{ for } 0 \le t < \pi;\ 1 + e^{-\pi s}/(s^2+1) \text{ for } t \ge \pi\} \Rightarrow$

$\mathcal{L}[Q] = \left\{\frac{1}{s^2+2s+2} \text{ for } 0 \le t < \pi;\ \frac{1}{s^2+2s+2} + \frac{e^{-\pi s}}{(s^2+1)(s^2+2s+2)} \text{ for } t \ge \pi\right\}$

$$\Rightarrow Q = \left\{ \mathcal{L}^{-1}\left[\frac{1}{s^2+2s+2}\right] \text{ for } 0 \le t < \pi;\ \mathcal{L}^{-1}\left[\frac{1}{s^2+2s+2} + \frac{e^{-\pi s}}{(s^2+1)(s^2+2s+2)}\right] \text{ for } t \ge \pi \right\}.$$

Now $\mathcal{L}^{-1}\left[\frac{1}{s^2+2s+2}\right] = \mathcal{L}^{-1}\left[\frac{1}{(s+1)^2+1}\right] = \mathcal{L}^{-1}\left[\frac{1}{(s-(-1))^2+1^2}\right] = e^{-t}\sin t$ and

$$\mathcal{L}^{-1}\left[\frac{1}{s^2+2s+2} + \frac{e^{-\pi s}}{(s^2+1)(s^2+2s+2)}\right] = e^{-t}\sin t + \mathcal{L}^{-1}\left[e^{-\pi s}\left(-\frac{1}{5}\cdot\frac{2s-1}{s^2+1} + \frac{1}{5}\cdot\frac{2s+3}{s^2+2s+2}\right)\right]$$

$= \text{[using a CAS]}\ e^{-t}\sin t + \frac{2}{5}\cos t - \frac{1}{5}\sin t - \frac{2}{5}e^{(\pi-t)}\cos t - \frac{1}{5}e^{(\pi-t)}\sin t.$

Therefore,

$$Q(t) = \begin{cases} e^{-t}\sin t & \text{for } 0 \le t < \pi \\ e^{-t}\sin t + \frac{2}{5}\cos t - \frac{1}{5}\sin t - \frac{2}{5}e^{(\pi-t)}\cos t - \frac{1}{5}e^{(\pi-t)}\sin t & \text{for } t \ge \pi \end{cases}.$$

17. $\mathcal{L}[f(t)] = 4\mathcal{L}[t] + \mathcal{L}[(f * \sin)(t)] \Rightarrow \mathcal{L}[f(t)] = \frac{4}{s^2} + \mathcal{L}[f(t)]\mathcal{L}[\sin(t)]$

$\Rightarrow \mathcal{L}[f(t)] = \frac{4}{s^2} + \frac{\mathcal{L}[f(t)]}{s^2+1} \Rightarrow \left(\frac{s^2}{s^2+1}\right)\mathcal{L}[f(t)] = \frac{4}{s^2} \Rightarrow \mathcal{L}[f(t)] = 4\cdot\frac{s^2+1}{s^4}$

$= 4\left(\frac{1}{s^2} + \frac{1}{s^4}\right) \Rightarrow f(t) = 4\left(\mathcal{L}^{-1}\left[\frac{1}{s^2}\right] + \mathcal{L}^{-1}\left[\frac{1}{s^4}\right]\right) = 4\left(t + \frac{1}{6}t^3\right) = 4t + \frac{2}{3}t^3.$

19. $\mathcal{L}[\dot{x}] = \mathcal{L}[1] - \mathcal{L}[(x * e^{-2t})(t)] = \frac{1}{s} - \mathcal{L}[x(t)]\mathcal{L}[e^{-2t}]$

$\Rightarrow s\mathcal{L}[x(t)] - x(0) = \frac{1}{s} - \mathcal{L}[x(t)]\cdot\frac{1}{s+2}$

$\Rightarrow \left(s + \frac{1}{s+2}\right)\mathcal{L}[x(t)] = 1 + \frac{1}{s} = \frac{s+1}{s} \Rightarrow \mathcal{L}[x(t)] = \frac{s+2}{s(s+1)} = \frac{1}{s+1} + 2\left(\frac{1}{s} - \frac{1}{s+1}\right)$

$= -\frac{1}{s+1} + \frac{2}{s} \Rightarrow x(t) = \mathcal{L}^{-1}\left[-\frac{1}{s+1} + \frac{2}{s}\right] = -\mathcal{L}^{-1}\left[\frac{1}{s+1}\right] + 2\mathcal{L}^{-1}\left[\frac{1}{s}\right] = 2 - e^{-t}.$

21. $\mathcal{L}[\dot{x}(t)] - 4\mathcal{L}[x(t)] + 4\mathcal{L}[(1 * x)(t)] = \mathcal{L}[t^3e^{2t}]$

$\Rightarrow s\mathcal{L}[x] - 4\mathcal{L}[x] + 4\mathcal{L}[1]\mathcal{L}[x] = \text{[entry 11 with } a = 2 \text{ and } f(t) = t^3]\ \frac{3!}{(s-2)^4}$

$\Rightarrow \left(s - 4 + \frac{4}{s}\right)\mathcal{L}[x] = \frac{6}{(s-2)^4} \Rightarrow \left(\frac{s^2-4s+4}{s}\right)\mathcal{L}[x] = \frac{6}{(s-2)^4}$

$\Rightarrow \mathcal{L}[x] = \frac{6s}{(s-2)^6} = 12\cdot\frac{1}{(s-2)^6} + 6\cdot\frac{1}{(s-2)^5}$

$\Rightarrow x(t) = 12\mathcal{L}^{-1}\left[\frac{1}{(s-2)^6}\right] + 6\mathcal{L}^{-1}\left[\frac{1}{(s-2)^5}\right] = 12\mathcal{L}^{-1}\left[\frac{1}{5!}\frac{5!}{(s-2)^6}\right] + 6\mathcal{L}^{-1}\left[\frac{1}{4!}\frac{4!}{(s-2)^5}\right]$

$= $ [using entry 11, with $a=2$ and $f(t)=t^5$ and t^4, resp.] $\frac{12}{5!}\left(t^5 e^{2t}\right) + \frac{6}{4!}\left(t^4 e^{2t}\right)$

$= \frac{1}{10}t^5 e^{2t} + \frac{1}{4}t^4 e^{2t}.$

6.3 Transforms of Discontinuous Functions

1. (a) The graph of $f(t)$ is

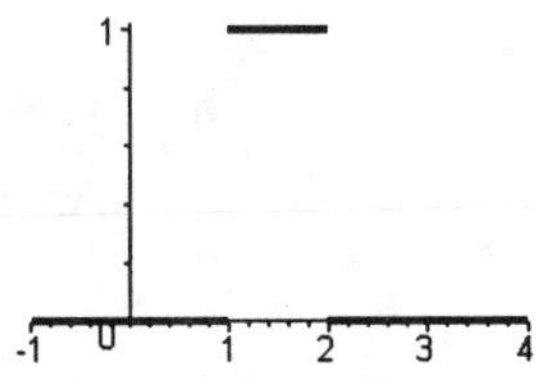

(b) $f(t) = 1\cdot U(t-1) + U(t-2)[0-1] = U(t-1) - U(t-2).$

3. (a) The graph of $f(t)$ is

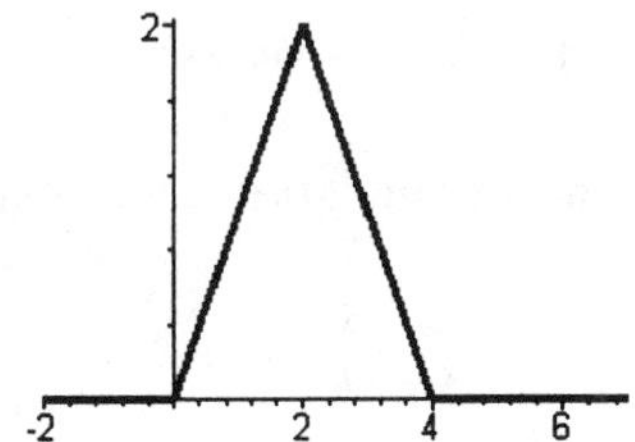

(b) $f(t) = t\cdot U(t) + U(t-2)[(4-t)-t] + U(t-4)[0-(4-t)] = tU(t) + (4-2t)U(t-2)$

$+(t-4)U(t-4).$

5. $\mathcal{L}[tU(t-a)] = \int_0^\infty tU(t-1)e^{-st}\,dt = \lim_{b\to\infty}\int_a^b te^{-st} = \lim_{b\to\infty}\left\{-\frac{1}{s}te^{-st}\Big|_a^b - \int_a^b\left(-\frac{1}{s}e^{-st}\right)dt\right\}$

$= \frac{ae^{-sa}}{s} + \lim_{b\to\infty}\frac{1}{s}\int_a^b e^{-st}\,dt = \frac{ase^{-sa}+e^{-sa}}{s^2} = (1+as)s^{-2}e^{-as}$ for $a>0$.

7. $\mathcal{L}[f(t)] = \mathcal{L}[U(t-1) - U(t-2)] = \mathcal{L}[U(t-1)] - \mathcal{L}[U(t-2)]$

$= $ [by (6.3.1a) with $f(t)\equiv 1$ and $a=1, 2$, respectively] $e^{-s}\mathcal{L}[1] - e^{-2s}\mathcal{L}[1]$

$= \frac{e^{-s}}{s} - \frac{e^{-2s}}{s} = \frac{e^{-s}\left(1-e^{-s}\right)}{s}.$

9. $\mathcal{L}[f(t)] = \mathcal{L}[tU(t) + (4-2t)U(t-2) + (t-4)U(t-4)] = \mathcal{L}[tU(t)] + \mathcal{L}[(4-2t)U(t-2)]$

$+\mathcal{L}[(t-4)U(t-4)] = \mathcal{L}[tU(t)] - 2\mathcal{L}[(t-2)U(t-2)] + \mathcal{L}[(t-4)U(t-4)]$

$= \dfrac{1}{s^2} - \dfrac{2e^{-2s}}{s^2} + \dfrac{e^{-4s}}{s^2} = \dfrac{1-2e^{-2s}+e^{-4s}}{s^2}.$

11. (a) We can express the equation as $P'(t) - kP(t) = -h[U(t) - U(t-30)]$. Then we have

$$[s\mathcal{L}[P] - P(0)] - k\mathcal{L}[P] = -\frac{h}{s} + \frac{he^{-30s}}{s} \Rightarrow (s-k)\mathcal{L}[P] = A - \frac{h}{s} + \frac{he^{-30s}}{s}$$

$\Rightarrow \mathcal{L}[P] = \dfrac{A}{s-k} - \dfrac{h}{s(s-k)} + \dfrac{he^{-30s}}{s(s-k)} \Rightarrow$ [using a CAS, although it can easily

be done by hand] $P(t) = \begin{cases} Ae^{kt} + \dfrac{h}{k}\left(1 - e^{kt}\right) & \text{for } 0 \le t \le 30 \\ Ae^{kt} + \dfrac{h}{k}\left(e^{-k(30-t)} - e^{kt}\right) & \text{for } t > 30. \end{cases}$

(b) We see that 330 days after the end of the 30-day fishing season, $t = 360$ and we want $P(360) = A$. Using the formula derived in part (a), we must have

$A = P(360) = Ae^{360k} + \dfrac{h}{k}\left(e^{-k(30-360)} - e^{360k}\right)$, or $\left(1 - e^{360k}\right)A = \dfrac{h}{k}\left(e^{-k(30-360)} - e^{360k}\right)$.

Therefore, $A = \dfrac{h}{k}\left(e^{330k} - e^{360k}\right) \Big/ \left(1 - e^{360k}\right)$.

13. We can write the equation as $4y' + 5y = (\sin 8t)U(t) - (\sin 8t)U(t-2)$. Therefore,

$4\mathcal{L}[y'] + 5\mathcal{L}[y] = \mathcal{L}[(\sin 8t)U(t)] - \mathcal{L}[(\sin 8t)U(t-2)]$

$=$ [using a CAS] $\dfrac{8 - 8e^{-2s}\cos(16) - se^{-2s}\sin(16)}{s^2+64} \Rightarrow 4[s\mathcal{L}[y] - y(0)] + 5\mathcal{L}[y]$

$= \dfrac{8 - 8e^{-2s}\cos(16) - se^{-2s}\sin(16)}{s^2+64} \Rightarrow (4s+5)\mathcal{L}[y] - 4 = \dfrac{8 - 8e^{-2s}\cos(16) - se^{-2s}\sin(16)}{s^2+64}$

$\Rightarrow \mathcal{L}[y] = \dfrac{264 - 8\cos(16)e^{-2s} - \sin(16)se^{-2s} + 4s^2}{(4s+5)(s^2+64)}$

$$\Rightarrow \text{[using a CAS]}\quad y(t)=\begin{cases}\frac{1081}{1049}e^{-5t/4}-\frac{32}{1049}\cos(8t)+\frac{5}{1049}\sin(8t), & t\le 2\\ e^{(-5t/4+5/2)}\left(\frac{5}{1049}\sin(16)-\frac{32}{1049}\cos(16)\right)+\frac{1081}{1049}e^{-5t/4}, & t>2\end{cases}$$

15. The equation can be written as $3y''+3y'+2y = 5[U(t)-U(t-5)]$. Therefore,

$$3[s^2\mathcal{L}[y]-s\,y(0)-y'(0)]+3[s\mathcal{L}[y]-y(0)]+2\mathcal{L}[y] = 5\left[\frac{1}{s}-\frac{e^{-5s}}{s}\right]$$

$$\Rightarrow (3s^2+3s+2)\mathcal{L}[y]=\frac{5}{s}-5e^{-5s}\cdot\frac{1}{s} \Rightarrow \mathcal{L}[y]=\frac{5}{s(3s^2+3s+2)}$$

$$-5e^{-5s}\frac{1}{s(3s^2+3s+2)} \Rightarrow y(t)=5\mathcal{L}^{-1}\left[\frac{1}{s(3s^2+3s+2)}\right]-5\mathcal{L}^{-1}\left[\frac{e^{-5s}}{s(3s^2+3s+2)}\right].$$

Now

$$\frac{1}{s(3s^2+3s+2)}=\frac{\frac{1}{2}}{s}-\frac{3}{2}\frac{s+1}{3s^2+3s+2}=\frac{\frac{1}{2}}{s}-\frac{3}{2}\left(\frac{1}{3}\frac{s+1}{(s+\frac{1}{2})^2+\left(\frac{\sqrt{15}}{6}\right)^2}\right)$$

$$=\frac{\frac{1}{2}}{s}-\frac{3}{2}\left\{\frac{1}{3}\left(\frac{s+\frac{1}{2}}{(s+\frac{1}{2})^2+\left(\frac{\sqrt{15}}{6}\right)^2}+\frac{\frac{1}{2}}{(s+\frac{1}{2})^2+\left(\frac{\sqrt{15}}{6}\right)^2}\right)\right\}$$

$$=\frac{\frac{1}{2}}{s}-\frac{1}{2}\left(\frac{s+\frac{1}{2}}{(s+\frac{1}{2})^2+\left(\frac{\sqrt{15}}{6}\right)^2}+\frac{\frac{1}{2}}{(s+\frac{1}{2})^2+\left(\frac{\sqrt{15}}{6}\right)^2}\right),\text{ so that}$$

$$5\mathcal{L}^{-1}\left[\frac{1}{s(3s^2+3s+2)}\right]=5\left(\frac{1}{2}-\frac{1}{2}e^{-t/2}\cos\left(\frac{\sqrt{15}\,t}{6}\right)-\frac{1}{4}\cdot\frac{6}{\sqrt{15}}\cdot e^{-t/2}\sin\left(\frac{\sqrt{15}\,t}{6}\right)\right)$$

$$=\frac{5}{2}-\frac{5}{2}e^{-t/2}\cos\left(\frac{\sqrt{15}\,t}{6}\right)-\frac{\sqrt{15}}{2}e^{-t/2}\sin\left(\frac{\sqrt{15}\,t}{6}\right)\quad\text{and}\quad 5\mathcal{L}^{-1}\left[\frac{e^{-5s}}{s(3s^2+3s+2)}\right]$$

$$=5\mathcal{L}^{-1}\left[e^{-5s}\left\{\frac{\frac{1}{2}}{s}-\frac{1}{2}\left(\frac{s+\frac{1}{2}}{(s+\frac{1}{2})^2+\left(\frac{\sqrt{15}}{6}\right)^2}+\frac{\frac{1}{2}}{(s+\frac{1}{2})^2+\left(\frac{\sqrt{15}}{6}\right)^2}\right)\right\}\right]$$

$$=5\left(\frac{1}{2}-\frac{1}{2}\cdot e^{-(t-5)/2}\cos\left(\frac{\sqrt{15}}{6}(t-5)\right)-\frac{1}{2}\cdot e^{-(t-5)/2}\sin\left(\frac{\sqrt{15}}{6}(t-5)\right)\right)U(t-5)$$

$$=\left(\frac{5}{2}-\frac{5}{2}\cdot e^{-(t-5)/2}\cos\left(\frac{\sqrt{15}}{6}(t-5)\right)-\frac{\sqrt{15}}{2}\cdot e^{-(t-5)/2}\sin\left(\frac{\sqrt{15}}{6}(t-5)\right)\right)U(t-5).$$

Therefore, $y(t) = \frac{5}{2}-\frac{1}{2}e^{-t/2}\left(5\cos\left(\frac{\sqrt{15}}{6}t\right)+\sqrt{15}\sin\left(\frac{\sqrt{15}}{6}t\right)\right)$ for $0\le t<5$;

$$y(t) = \frac{1}{2}e^{(-t+5)/2}\left(5\cos\left(\frac{\sqrt{15}}{6}(t-5)\right)+\sqrt{15}\sin\left(\frac{\sqrt{15}}{6}(t-5)\right)\right)$$
$$-\frac{1}{2}e^{-t/2}\left(5\cos\left(\frac{\sqrt{15}}{6}t\right)+\sqrt{15}\sin\left(\frac{\sqrt{15}}{6}t\right)\right)\quad\text{for } t\ge 5.$$

17. The function whose graph is given can be expressed as $g(t)=t$ for $0\le t\le 1$; $t-1$ for $1\le t\le 2$; 0 elsewhere. In terms of the unit step function, we can write

$g(t)=tU(t)-U(t-1)+(1-t)U(t-2)=tU(t)-U(t-1)-(t-1)U(t-2)$

$=tU(t)-U(t-1)-(t-2)U(t-2)-U(t-2)$. Applying the Laplace transform to each side of the differential equation, we get

$$(s+1)\mathcal{L}[y]=\frac{1}{s^2}-\frac{e^{-s}}{s}-\frac{e^{-2s}}{s^2}-\frac{e^{-2s}}{s}\Rightarrow\mathcal{L}[y]=\frac{1}{s^2(s+1)}-\frac{e^{-s}}{s(s+1)}-\frac{e^{-2s}}{s^2(s+1)}-\frac{e^{-2s}}{s(s+1)}$$

$\Rightarrow y(t)=t-1+e^{-t}+\left(e^{1-t}-1\right)U(t-1)+(2-t)U(t-2)$, which can be expressed as

$$y(t)=\begin{cases} t-1+e^{-t} & \text{for } 0\le t\le 1\\ t-2+e^{-t}+e^{(1-t)} & \text{for } 1\le t\le 2\\ e^{-t}+e^{(1-t)} & \text{for } t>2.\end{cases}$$

19. (a) The graph of $W(t)$ is

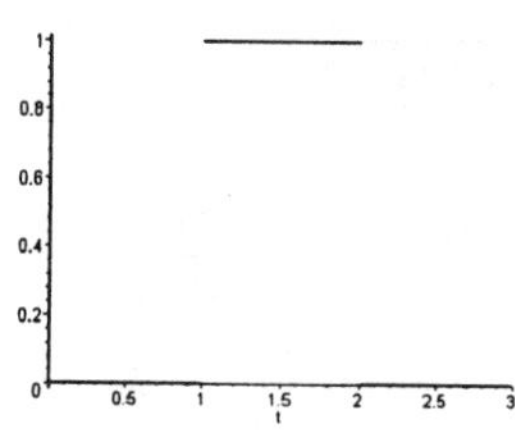

(b) $y''+3y'+2y=0\Rightarrow\lambda^2+3\lambda+2=(\lambda+2)(\lambda+1)=0$
$\Rightarrow y(t)=c_1e^{-2t}+c_2e^{-t}$. Using the initial conditions we get the algebraic system $\{c_1+c_2=0,\ -2c_1-c_2=0\}$, which has the solution $c_1=c_2=0$. This means that $y(t)\equiv 0$.

(c) We have $\mathcal{L}[y'']+3\mathcal{L}[y']+2\mathcal{L}[y]=\mathcal{L}[W(t)]=\mathcal{L}[U(t-1)]-\mathcal{L}[U(t-2)]$

$$\Rightarrow\left(s^2+3s+2\right)\mathcal{L}[y]=\frac{e^{-s}}{s}-\frac{e^{-2s}}{s}\Rightarrow\mathcal{L}[y]=\frac{e^{-s}}{s\left(s^2+3s+2\right)}-\frac{e^{-2s}}{s\left(s^2+3s+2\right)}$$

$$\Rightarrow y(t)=\mathcal{L}^{-1}\left[e^{-s}\cdot\left(\frac{\frac{1}{2}}{s}-\frac{1}{s+1}+\frac{\frac{1}{2}}{s+2}\right)\right]-\mathcal{L}^{-1}\left[e^{-2s}\cdot\left(\frac{\frac{1}{2}}{s}-\frac{1}{s+1}+\frac{\frac{1}{2}}{s+2}\right)\right]$$

$$=\left(\frac{1}{2}-e^{-(t-1)}+\frac{1}{2}e^{-2(t-1)}\right)U(t-1)-\left(\frac{1}{2}-e^{-(t-2)}+\frac{1}{2}e^{-2(t-2)}\right)U(t-2)$$

$$= \begin{cases} 0 & \text{for } 0 \le t < 1 \\ \frac{1}{2}e^{2(1-t)} + \frac{1}{2} - e^{1-t} & \text{for } 1 \le t \le 2 \\ \frac{1}{2}e^{2(1-t)} - e^{1-t} + e^{2-t} - \frac{1}{2}e^{2(2-t)} & \text{for } t > 2. \end{cases}$$

(d) The graphs of the solutions to parts (b) and (c) are

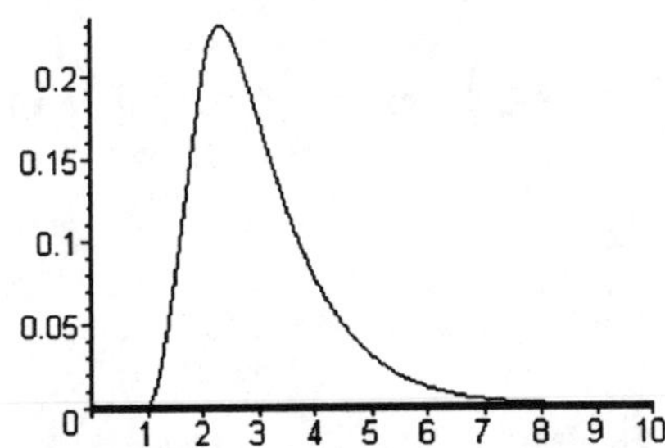

The forcing term creates a temporary "blip" in the zero function. However, the solution begins decaying exponentially shortly after 2 units of time.

6.4 Transforms of Impulse Functions—the Dirac Delta Function

1. $\mathcal{L}[y''] = \mathcal{L}[\delta(t\text{-}a)] \Rightarrow [s^2\mathcal{L}[y] - sy(0) - y'(0)] = e^{-sa} \Rightarrow \mathcal{L}[y] = e^{-sa} \cdot \dfrac{1}{s^2} = e^{-sa}\mathcal{L}[t]$

$\Rightarrow y = \mathcal{L}^{-1}\left[e^{-sa}\mathcal{L}[t]\right] =$ [by (6.3.1b)] $(t-a)U(t-a)$

$$= \begin{cases} 0 & \text{for } t < a \\ t-a & \text{for } t \ge a. \end{cases}$$

3. $2\mathcal{L}[y''] + \mathcal{L}[y'] + 2\mathcal{L}[y] = \mathcal{L}[\delta(t-5)] \Rightarrow 2[s^2\mathcal{L}[y] - sy(0) - y'(0)] + [s\mathcal{L}[y] - y(0)] + 2\mathcal{L}[y]$

$= e^{-5s} \Rightarrow (2s^2 + s + 2)\mathcal{L}[y] = e^{-5s} \Rightarrow \mathcal{L}[y] = e^{-5s} \cdot \dfrac{1}{2s^2 + s + 2} = e^{-5s} \cdot \dfrac{1}{2} \cdot \dfrac{1}{\left(s + \frac{1}{4}\right)^2 + \left(\frac{\sqrt{15}}{4}\right)^2}$

$= e^{-5s} \cdot \dfrac{1}{2} \cdot \dfrac{1}{\frac{\sqrt{15}}{4}} \cdot \dfrac{\frac{\sqrt{15}}{4}}{\left(s + \frac{1}{4}\right)^2 + \left(\frac{\sqrt{15}}{4}\right)^2} = \dfrac{2\sqrt{15}}{15} e^{-5s} \mathcal{L}\left[e^{-t/4} \sin\left(\dfrac{\sqrt{15}}{4} t\right)\right]$

$\Rightarrow y(t) = \dfrac{2\sqrt{15}}{15} e^{-(t-5)/4} \sin\left(\dfrac{\sqrt{15}}{4}(t-5)\right) U(t-5)$

$$= \begin{cases} 0 & \text{for } t < 5 \\ \dfrac{2\sqrt{15}}{15} e^{-(t-5)/4} \sin\left(\dfrac{\sqrt{15}}{4}(t-5)\right) & \text{for } t \ge 5. \end{cases}$$

5. $\mathcal{L}[y''] + 6\mathcal{L}[y'] + 109\mathcal{L}[y] = \mathcal{L}[\delta(t-1)] - \mathcal{L}[\delta(t-7)] \Rightarrow [s^2\mathcal{L}[y] - sy(0) - y'(0)]$

$+ 6[s\mathcal{L}[y] - y(0)] + 109\mathcal{L}[y] = e^{-s} - e^{-7s} \Rightarrow (s^2 + 6s + 109)\mathcal{L}[y] = e^{-s} - e^{-7s}$

$$\Rightarrow \mathcal{L}[y]=\frac{e^{-s}}{(s+3)^2+10^2} - \frac{e^{-7s}}{(s+3)^2+10^2}$$

$$\Rightarrow \text{[by (6.3.1b)]}\ y(t)=\left[\tfrac{1}{10}e^{-3(t-1)}\sin(10(t-1))\right]U(t-1) - \left[\tfrac{1}{10}e^{-3(t-7)}\sin(10(t-7))\right]U(t-7).$$

7. $\mathcal{L}[y'']+\mathcal{L}[y]=4\mathcal{L}\left[\delta\left(t-\frac{3\pi}{2}\right)\right]\Rightarrow (s^2+1)\mathcal{L}[y]-1=4e^{-\frac{3}{2}\pi s} \Rightarrow \mathcal{L}[y]=\frac{1}{s^2+1}+\frac{4e^{-\frac{3}{2}\pi s}}{s^2+1}$

$$\Rightarrow y(t)=\sin t \ +\ 4\sin\left(t-\tfrac{3\pi}{2}\right)U\left(t-\tfrac{3\pi}{2}\right)=\sin t \ +\ 4\cos(t)U\left(t-\tfrac{3\pi}{2}\right)$$

$$=\begin{cases}\sin t & \text{for } t<3\pi/2\\ \sin t+4\cos t & \text{for } t\geq 3\pi/2.\end{cases}$$

9. $EI\mathcal{L}\left[y^{(iv)}\right] = W\mathcal{L}\left[\delta\left(x-\frac{L}{2}\right)\right] \Rightarrow EI\left[s^4\mathcal{L}[y]-s^3y(0)-s^2y'(0)-sy''(0)-y'''(0)\right]=We^{-\frac{L}{2}s}$

$\Rightarrow EI\left[s^4\mathcal{L}[y]-sy''(0)-y'''(0)\right]=We^{-\frac{L}{2}s} \Rightarrow$ [letting $C_1=y''(0)$, $C_2=y'''(0)$ as in Ex. 6.3.2]

$$\Rightarrow \mathcal{L}[y]=\frac{C_1}{s^3}+\frac{C_2}{s^4}+\frac{W}{EI}\cdot e^{-\frac{L}{2}s}\cdot\frac{1}{s^4} \Rightarrow y(x)=\frac{C_1}{2}\cdot x^2+\frac{C_2}{6}\cdot x^3+\frac{W}{6EI}\cdot\left(x-\frac{L}{2}\right)^3\cdot U\left(x-\frac{L}{2}\right).$$

Now $y''(L)=0 \Rightarrow 0=C_1+LC_2+\frac{WL}{2EI}$ and $y'''(L)=0 \Rightarrow 0=C_2+\frac{W}{EI} \Rightarrow C_2=-\frac{W}{EI}$

$\Rightarrow$ [using our first boundary condition] $C_1=\frac{WL}{2EI}$

$$\Rightarrow y(x)= \text{[after expanding and simplifying]} \begin{cases}\frac{W}{6EI}x^2\left(\frac{3}{2}L-x\right) & \text{for } 0\leq x<\frac{L}{2}\\ \frac{WL^2}{24EI}\left(3x-\frac{L}{2}\right) & \text{for } x\geq\frac{L}{2}.\end{cases}$$

11. $\mathcal{L}[y'']+a\mathcal{L}[y']+b\mathcal{L}[y]=\mathcal{L}[f(t)]\Rightarrow (s^2+as+b)\mathcal{L}[y]-s\,y(0)-y'(0)$

$=\mathcal{L}[f(t)]\Rightarrow \mathcal{L}[y]=\frac{\mathcal{L}[f(t)]+s\,y(0)+y'(0)}{s^2+as+b}$. If $f(t)$ is replaced by $f(t)+c\delta(t)$ on the right side of the equation, the transformed solution is given by the equation $\mathcal{L}[y]=\frac{\mathcal{L}[f(t)]+s\,y(0)+y'(0)+c}{s^2+as+b}$, so the effect is that we have increased the initial value of $y'(0)$ by the constant c. (This is a consequence of the fact that $\mathcal{L}[\delta(t)]=1$.)

13. (a) $\mathcal{L}[\delta(t-a)f(t)] = \int_0^\infty \delta(t-a)f(t)e^{-st}\,dt = \int_0^\infty \delta(t-a)\left[f(t)e^{-st}\right]dt$

$= f(a)e^{-as}$ by the result of Problem 12, with $g(t)=f(t)e^{-st}$.

(b) $\mathcal{L}[y'']+2\mathcal{L}[y']+\mathcal{L}[y]=\mathcal{L}[\delta(t-1)t]\Rightarrow \left[s^2\mathcal{L}[y]-sy(0)-y'(0)\right]+2\left[s\mathcal{L}[y]-y(0)\right]+\mathcal{L}[y]$

$=e^{-1\cdot s}\cdot 1=e^{-s} \Rightarrow (s^2+2s+1)\mathcal{L}[y] = e^{-s} \Rightarrow \mathcal{L}[y] = e^{-s}\cdot\frac{1}{(s+1)^2}$

$$\Rightarrow y(t) = (t-1)e^{-(t-1)}U(t-1) = \begin{cases} 0 & \text{for } 0 \le t < 1 \\ (t-1)e^{-(t-1)} & \text{for } t \ge 1. \end{cases}$$

6.5 Transforms and Systems of Linear Differential Equations

1. We have (1) $(s+3)\mathcal{L}[x(t)] - \mathcal{L}[y(t)] = 2$ and (2) $(s+3)\mathcal{L}[y(t)] - \mathcal{L}[x(t)] = 3$. Multiplying equation (2) by $(s+3)$ and then adding (1) and the new equation (2), we get $(s+3)^2\ \mathcal{L}[y(t)] - \mathcal{L}[y(t)] = 2 + 3(s+3)$, or $(s^2+6s+8)\mathcal{L}[y(t)] = 3s+11$, so that
$$\mathcal{L}[y(t)] = \frac{3s+11}{(s^2+6s+8)} = \frac{3s+11}{(s+2)(s+4)} = \frac{\frac{5}{2}}{s+2} + \frac{\frac{1}{2}}{s+4}$$
$$\Rightarrow y(t) = \frac{5}{2}e^{-2t} + \frac{1}{2}e^{-4t}.$$

3. We have $\mathcal{L}[x'] = [s\mathcal{L}[x] - x(0)] = 12\mathcal{L}[x] + 5\mathcal{L}[y]$ and $\mathcal{L}[y'] = [s\mathcal{L}[y] - y(0)] = -6\mathcal{L}[x] + \mathcal{L}[y]$ or, letting $\mathcal{L}[x] = X$ and $\mathcal{L}[y] = Y$, (1) $(s-12)X - 5Y = 0$ and (2) $6X + (s-1)Y = 1$. Now $(s-1)\times(1) + 5\times(2)$ yields $(s^2 - 13s + 42)X = 5$, or $X = \dfrac{5}{(s-6)(s-7)} = \dfrac{5}{s-7} - \dfrac{5}{s-6}$. Therefore, $x(t) = \mathcal{L}^{-1}[X]$ $= 5e^{7t} - 5e^{6t}$. Then the first ODE implies that $y(t) = \frac{1}{5}(x' - 12x) = -5e^{7t} + 6e^{6t}$.

5. $\mathcal{L}[x'] = [s\mathcal{L}[x] - x(0)] = -6\mathcal{L}[x] + 2\mathcal{L}[y]$ and $\mathcal{L}[y'] = [s\mathcal{L}[y] - y(0)]$ $= -7\mathcal{L}[x] + 3\mathcal{L}[y]$. This system can be written as (1) $(s+6)X - 2Y = 1$ and (2) $7X + (s-3)Y = 0$. Now $(s-3)\times(1) + 2\times(2) \Rightarrow (s^2 + 3s - 4)X = s - 3$
$\Rightarrow X = \dfrac{s-3}{(s+4)(s-1)} = \dfrac{\frac{7}{5}}{s+4} - \dfrac{\frac{2}{5}}{s-1} \Rightarrow x(t) = \frac{7}{5}e^{-4t} - \frac{2}{5}e^{t}$. Then the first differential equation in the system implies that $y(t) = \frac{1}{2}(x' + 6x) = \frac{7}{5}e^{-4t} - \frac{7}{5}e^{t}$.

7. $\mathcal{L}[x'] + \mathcal{L}[y'] = -3\mathcal{L}[x] - 2\mathcal{L}[y] + \mathcal{L}[e^{-2t}]$ and $2\mathcal{L}[x'] + \mathcal{L}[y'] = -2\mathcal{L}[x] - \mathcal{L}[y]$ $+\mathcal{L}[1]$. This system can be written as (1) $(s+3)X + (s+2)Y = \dfrac{1}{s+2}$ and (2) $2(s+1)X + (s+1)Y = \dfrac{1}{s}$. Now $(s+1)\times(1) + -(s+2)\times(2) \Rightarrow -(s+1)^2 X$ $= -\dfrac{3s+4}{s(s+2)} \Rightarrow$ [by partial fractions] $X = \dfrac{2}{s} - \dfrac{1}{(s+2)^2} - \dfrac{3}{s+1} + \dfrac{1}{s+2}$ and $x(t) = 2 - te^{-t} - 3e^{-t} + e^{-2t}$. Then equation (2) implies that $Y = \left\{\dfrac{1}{s} - 2(s+1)X\right\}\Big/(s+1)$ $= -\dfrac{3}{s} + \dfrac{2}{(s+1)^2} + \dfrac{5}{s+1} - \dfrac{2}{s+2}$, so we have $y(t) = -3 + 2te^{-t} + 5e^{-t} - 2e^{-2t}$.

9. $\mathcal{L}[x'] + \mathcal{L}[y'] = \mathcal{L}[x]$, $\mathcal{L}[y'] + \mathcal{L}[z'] = \mathcal{L}[x]$, and $\mathcal{L}[z'] + \mathcal{L}[x'] = \mathcal{L}[x]$ $\Rightarrow$ (1) $(s-1)X + sY = 2$, (2) $-X + sY + sZ = 2$, (3) $(s-1)X + sZ = 2$. Now (1) − (3) + (2) $\Rightarrow -X + 2sY = 2$. Adding $-2\times(1)$ to this last equation, we get $(-2s+1)X = -2$, which implies $X = \dfrac{2}{2s-1} = \dfrac{1}{s-\frac{1}{2}}$. Therefore, $x(t) = e^{t/2}$. Now equation (1) $\Rightarrow y' = x - x'$ $= \frac{1}{2}e^{t/2} \Rightarrow y(t) = e^{t/2} + C$, and $y(0) = 1 \Rightarrow C = 0$, so $y(t) = e^{t/2}$. Finally, the third differential equation gives us $z' = x - x'$, so that $z(t) = e^{t/2}$ in the same way we found $y(t)$.

11. $\mathcal{L}[x''] - \mathcal{L}[y'] = -\mathcal{L}[t] + \mathcal{L}[1]$ and $\mathcal{L}[x'] - \mathcal{L}[x] + 2\mathcal{L}[y'] = 4\mathcal{L}[e^t]$ $\Rightarrow$ (1) $s^2X - sY = 1 + \dfrac{s-1}{s^2}$ and (2) $(s-1)X + 2sY = \dfrac{4}{s-1}$. Then $2\times(1) + (2)$ $\Rightarrow (2s-1)(s+1)X = 2\left(1+\dfrac{s-1}{s^2}\right) + \dfrac{4}{s-1} \Rightarrow X = \dfrac{2}{(2s-1)(s+1)} + \dfrac{2(s-1)}{s^2(2s-1)(s+1)}$ $+ \dfrac{4}{(2s-1)(s+1)(s-1)} = -\dfrac{10}{3}\dfrac{1}{s-\frac{1}{2}} + \dfrac{\frac{4}{3}}{s+1} + \dfrac{2}{s^2} + \dfrac{2}{s-1} \Rightarrow x(t) = -\dfrac{10}{3}e^{t/2} + \dfrac{4}{3}e^{-t}$ $+ 2t + 2e^t$. Now the second differential equation indicates that $y' = \frac{1}{2}\{4e^t + x - x'\}$ $= 2e^t - \dfrac{5}{6}e^{t/2} + \dfrac{4}{3}e^{-t} + t - 1$, so $y(t) = 2e^t - \dfrac{5}{3}e^{t/2} - \dfrac{4}{3}e^{-t} + \dfrac{1}{2}t^2 - t + C$. The initial condition $y(0) = 0$ implies that $C = 1$ and so $y(t) = 2e^t - \dfrac{5}{3}e^{t/2} - \dfrac{4}{3}e^{-t} + \dfrac{1}{2}t^2 - t + 1$.

13. We have the system $x'' = -(x-\theta) - 2(x+\theta)$, $\theta'' = (x-\theta) - 2(x+\theta)$; $x(0) = 1$, $x'(0) = 0$, $\theta(0) = 0.1$, $\theta'(0) = 0$. Using a CAS (*Maple* in this case, with *method* = *laplace* as an option in *dsolve*), we find that $x(t) = \dfrac{11}{20}\cos 2t + \dfrac{9}{20}\cos\left(\sqrt{2}\,t\right)$ and $\theta(t) = \dfrac{11}{20}\cos 2t - \dfrac{9}{20}\cos\left(\sqrt{2}\,t\right)$.

15. The system is $2\dot{I}_1 + 3(I_1 - I_2) = 6$, $2\dot{I}_2 + 8I_2 + 3(I_2 - I_1) = 0$; $I_1(0) = I_2(0) = 0$. Using a CAS, we find that $I_1(t) = \dfrac{11}{4} - \dfrac{1}{20}e^{-6t} - \dfrac{27}{10}e^{-t}$ and $I_2(t) = \dfrac{3}{4} + \dfrac{3}{20}e^{-6t} - \dfrac{9}{10}e^{-t}$.

6.6 Qualitative Analysis Via the Laplace Transform

1. $3s+5=0 \Rightarrow s=-5/3 \Rightarrow x(t)\to 0$ as $t\to\infty$.

3. $s^2+2s+10=0 \Rightarrow s=-1\pm 3i \Rightarrow x(t)\to 0$ as $t\to\infty$ because the real parts of the poles are negative. There are oscillations, but the amplitudes are diminishing.

5. $s^2+6s+18=0 \Rightarrow s=-3\pm 3i \Rightarrow x(t)\to 0$ as $t\to\infty$. There are oscillations with decreasing amplitudes.

7. $s^4+5s^2+4=0 \Rightarrow s=\pm 2i, \pm i \Rightarrow x(t)$ oscillates as $t\to\infty$.

9. (a) $\mathcal{L}[\ddot{x}]+2\mathcal{L}[\dot{x}]+2\mathcal{L}[x]=\mathcal{L}\left[e^{-t/10}\right]\Rightarrow\left[s^2X-s\,x(0)-x'(0)\right]+2\left[sX-x(0)\right]+2X=\dfrac{1}{s+\frac{1}{10}}$

$$=\frac{10}{10s+1}\Rightarrow\left(s^2+2s+2\right)X=4s+9+\frac{10}{10s+1}\Rightarrow X=\frac{40s^2+94s+19}{(10s+1)\left(s^2+2s+2\right)}.$$

(b) From part (a), the poles are $s=-1/10$ and $s=-1\pm i$.

(c) Because the real pole is negative and the real part of the complex pole is also negative, we see that $x(t)$ tends to 0 as t grows large. There are oscillations, but the amplitudes are diminishing.

11. (a) $2\mathcal{L}[\ddot{x}]+7\mathcal{L}[\dot{x}]+3\mathcal{L}[x]=\mathcal{L}[2\cos t]\Rightarrow 2\left[s^2X-s\,x(0)-\dot{x}(0)\right]+7\left[sX-x(0)\right]+3X$

$$=\frac{2s}{s^2+1}\Rightarrow\left(2s^2+7s+3\right)X=2s+7+\frac{2s}{s^2+1}\Rightarrow X=\frac{2s^3+7s^2+4s+7}{\left(s^2+1\right)\left(2s^2+7s+3\right)}.$$

(b) The poles are $s=-1/2,-3,i$, and $-i$.

(c) The poles indicate that there are two transient terms and two oscillating terms, so the solution becomes oscillatory as t grows larger.

13. If x is the unique solution of the modified IVP $c_2x''+c_1x'+c_0x=f(t),\ x(0)=0=x'(0)$, then the transfer function $R=\dfrac{X(s)}{F(s)}$ gives us $X(s)=R(s)\,F(s)$, and the Convolution Theorem (Section 6.2) yields $x(t)=\mathcal{L}^{-1}[X(s)]=\mathcal{L}^{-1}[R(s)F(s)]=(r*f)(t)$, where $r=\mathcal{L}^{-1}\{R\}(t)$ is the response function. This says that the convolution $(r*f)$ is the (unique) solution of the (modified) IVP. Now let x_H be the unique solution of the homogeneous equation $c_2x''+c_1x'+c_0x=0$, with $x(0)=x_0$ and $x'(0)=x_1$. [The first Existence and Uniqueness Theorem of Section 4.6 guarantees uniqueness of all solutions in this exercise.] Then $x_{GNH}=x_{GH}+x_{PNH}=x_H+(r*f)$. Furthermore, $x_{GNH}(0)=x_H(0)+(r*f)(0)=x_0+0=x_0$ and $x'_{GNH}(0)=x'_H(0)+(r*f)'(0)=x_1+0=x_1$, and so $x_{GNH}(t)=x_H(t)+(r*f)(t)$ is the unique solution of the original IVP.

15. (a) Assuming that $y(0)=y'(0)=0$ and using equation (6.6.5), we see that the transfer function is $\frac{Y(s)}{G(s)}=\frac{1}{s^2-s-6}$.

(b) The impulse response function is the inverse Laplace transform of the transfer function found in part (a), so we want $\mathcal{L}^{-1}\left[\frac{1}{s^2-s-6}\right]=\mathcal{L}^{-1}\left[\frac{1}{(s-3)(s+2)}\right]=\mathcal{L}^{-1}\left[\frac{1}{5}\left(\frac{1}{s-3}-\frac{1}{s+2}\right)\right]$ $=\frac{1}{5}\left\{\mathcal{L}^{-1}\left[\frac{1}{s-3}\right]-\mathcal{L}^{-1}\left[\frac{1}{s+2}\right]\right\}=\frac{1}{5}\left\{e^{3t}-e^{-2t}\right\}$.

(c) First we find the unique solution of the homogeneous IVP $y''-y'-6y=0;\ y(0)=1,\ y'(0)=8$. The characteristic equation is $\lambda^2-\lambda-6=(\lambda-3)(\lambda+2)=0$, giving us the general solution $y(t)=c_1e^{3t}+c_2e^{-2t}$. The initial conditions imply that $c_1=2$ and $c_2=-1$, so that $y_{\mathrm{H}}=2e^{3t}-e^{-2t}$. Now using the result of Exercise 13, with $R=\frac{1}{s^2-s-6}$ and $r=\mathcal{L}^{-1}\{R\}(t)=\frac{1}{5}e^{3t}-\frac{1}{5}e^{-2t}$, we find that the solution of the IVP is $y(t)=\int_0^t\left\{\frac{1}{5}e^{3(t-u)}-\frac{1}{5}e^{-2(t-u)}\right\}g(u)\,du+2e^{3t}-e^{-2t}=\frac{1}{5}e^{3t}\int_0^t e^{-3u}g(u)\,du$ $-\frac{1}{5}e^{-2t}\int_0^t e^{2u}g(u)\,du+2e^{3t}-e^{-2t}$.

17. (a) Assuming that $x(0)=0$, we apply the Laplace transform to each side of the equation to find that the transfer function is $\frac{X(s)}{F(s)}=\frac{1}{a_1s+a_0}$.

(b) First note that if we solve the associated homogeneous equation (with no initial condition), we get $x_{\mathrm{GH}}=ke^{-(a_0/a_1)t}$. The transfer function found in part (a) can be written as $W(s)=\frac{1}{a_1s+a_0}=\frac{1/a_0}{1+(a_1/a_0)s}=\frac{c}{1+Ts}$, where $T=a_1/a_0$ is called the *time constant*. [When a transient term decays because of an exponential term of the form e^{-rt}, the number $T=1/r$, which characterizes the rate of decay of the transient term, is called the *time constant*. When $t=1/r$, we see that $e^{-rt}=e^{-1}\approx 0.37$, which means that the transient term has decayed to a little more than one-third its original size.]

Chapter 7

Systems of Nonlinear Differential Equations

7.1 Equilibria of Nonlinear Systems

1. $(1)\,-x+xy=0$, $(2)\,-y+2xy=0$: $(1)\Rightarrow x(y-1)=0\Rightarrow x=0$ or $y=1$. Now $x=0$ $\Rightarrow$ [using (2)] $y=0$; and $y=1\Rightarrow$ [using (2)] $x=\frac{1}{2}$. Therefore, the equilibrium points are $(0,\,0)$ and $\left(\frac{1}{2},\,1\right)$.

3. $(1)\,x^2-y^2=0$, $(2)\,x-xy=0$: $(2)\Rightarrow x(1-y)=0\Rightarrow x=0$ or $y=1$. Now $x=0$ $\Rightarrow$ [using (1)] $y=0$; and $y=1\Rightarrow$ [using (1)] $x=\pm 1$. Therefore, the equilibrium points are $(0,\,0)$, $(1,\,1)$, and $(-1,\,1)$.

5. $(1)\,x+y+2xy=0$, $(2)\,-2x+y+y^3=0$: $(1)\Rightarrow x=-y/(1+2y)$; substituting for x in (2) $\Rightarrow 3y+2y^2+y^3+2y^4=y(3+2y+y^2+2y^3)=0\Rightarrow y=0$ or $3+2y+y^2+2y^3=0$. Now $y=0\Rightarrow$ [using (1) or (2)] $x=0$; and $3+2y+y^2+2y^3=0\Rightarrow$ [using technology if necessary] $y=-1$ is the only real solution $\Rightarrow$ [using (1) or (2)] $x=-1$. Therefore, the equilibrium points are $(0,\,0)$ and $(-1,\,-1)$.

7. $(1)\,x-x^2-xy=0$, $(2)\,3y-xy-2y^2=0$: $(1)\Rightarrow x(1-x-y)\Rightarrow x=0$ or $x+y=1$. Now $x=0\Rightarrow$ [using (2)] $3y-2y^2=y(3-2y)=0\Rightarrow y=0$ or $y=\frac{3}{2}$. On the other hand, substituting $x+y=1$ in (2) $\Rightarrow 2y-y^2=y(2-y)=0\Rightarrow y=0$ or $y=2$. But then, because $x+y=1$, we have $x=1$ or $x=-1$. Putting the pieces together, we have the equilibrium points $(0,\,0)$, $\left(0,\,\frac{3}{2}\right)$, $(1,\,0)$, and $(-1,\,2)$.

9. $(1)\,(1+x)\sin y=0$, $(2)\,1-x-\cos y=0$: $(1)\Rightarrow x=-1$ or $y=n\pi$ $(n=0,\,\pm 1,\pm 2,\ldots)$. Now $x=-1\Rightarrow$ [using (2)] $\cos y=2$, an impossibility. Then $y=n\pi\Rightarrow$ [using (2)] $x=0$ for even multiples of π and $x=2$ for odd multiples of π. Therefore, the equilibrium points are $(0,\,2n\pi)$ and $(2,\,(2n+1)\pi)$, $n=0,\,\pm 1,\,\pm 2,\ldots$ The graphical method suggested in the statement of the problem may not lead to enlightenment.

11. In *Maple*, the following commands give the equilibrium points (0, 0), $\left(0, -\dfrac{k_2 - G_2N_0}{a_2G_2}\right)$, and $\left(-\dfrac{k_1 - G_1N_0}{a_1G_1}, 0\right)$:

```
> N:=N0-a1*n1-a2*n2:
> solve({G1*N*n1-k1*n1=0,G2*N*n2-k2*n2=0},{n1,n2});
```

$$\{n2 = 0, n1 = 0\}, \{n2 = -\frac{-G2\ N0 + k2}{G2\ a2}, n1 = 0\}, \{n2 = 0, n1 = \frac{G1\ N0 - k1}{G1\ a1}\}$$

13. (a) $\dfrac{d^2\theta}{dt^2} + \dfrac{g}{L}\sin\theta$: Then $x_1 = \theta$ and $x_2 = \dfrac{d\theta}{dt} \Rightarrow \dfrac{dx_1}{dt} = x_2$ and $\dfrac{dx_2}{dt} = \dfrac{d^2\theta}{dt^2} = -\dfrac{g}{L}\sin\theta$ $= -\dfrac{g}{L}\sin x_1$.

(b) The equilibrium points must satisfy $(1)\ x_2 = 0$ and $(2) -\dfrac{g}{L}\sin x_1 = 0$. Equation (2) $\Rightarrow x_1 = n\pi\ (n = 0, \pm1, \pm2, \ldots)$. Therefore, the equilibrium points of the system are $(n\pi, 0)$, $n = 0, \pm1, \pm2, \ldots$. **[Note that there is a typo in the answer given at the back of the text.]**

7.2 Linear Approximation at Equilibrium Points

1. (a) $x^2 + y^2 = r^2 \Rightarrow \dfrac{d}{dt}(x^2 + y^2) = \dfrac{d}{dt}r^2 \Rightarrow$ [by the Chain Rule] $= 2x\dot{x} + 2y\dot{y} = 2r\dot{r}$, or $x\dot{x} + y\dot{y} = r\dot{r}$.

(b) Substituting for $\dot{x}$ and $\dot{y}$ in the expression for $r\dot{r}$ derived in part (a), we have $r\dot{r} = x[y + ax(x^2 + y^2)] + y[-x + ay(x^2 + y^2)] = ax^2(x^2 + y^2) + ay^2(x^2 + y^2)$ $= a(x^2 + y^2)^2 = ar^4$. Dividing by $r > 0$, we get $\dot{r} = ar^3$.

(c) Tan $\theta = \dfrac{r\sin\theta}{r\cos\theta} = \dfrac{y}{x} \Rightarrow \theta = \arctan(y/x)$. Then $\dot{\theta} =$ [Chain Rule] $\dfrac{1}{1 + (y/x)^2} \cdot \left(\dfrac{x\dot{y} - \dot{x}y}{x^2}\right)$ $= \dfrac{x^2}{x^2 + y^2}\left(\dfrac{x\dot{y} - \dot{x}y}{x^2}\right) = \dfrac{x\dot{y} - \dot{x}y}{x^2 + y^2} = \dfrac{x\dot{y} - \dot{x}y}{r^2}$. Substituting the differential equations from the system into this last expression, we find that $\dot{\theta} = \dfrac{x\dot{y} - \dot{x}y}{r^2}$

$$= \frac{x[-x + ay(x^2 + y^2)] - [y + ax(x^2 + y^2)]y}{r^2} = \frac{-(x^2 + y^2)}{r^2} = -1.$$

(d) If $a < 0$, then $\dot{r} < 0$, implying that a trajectory will spiral into (0, 0), so that the origin is a *stable spiral point* (a sink). If $a = 0$, then $\dot{r} = 0$, so that r is a constant and the origin is a *stable center*. If $a > 0$, the origin is an *unstable spiral point* (a source). In the language of Section 2.6, the parameter value $a = 0$ is a *bifurcation point*.

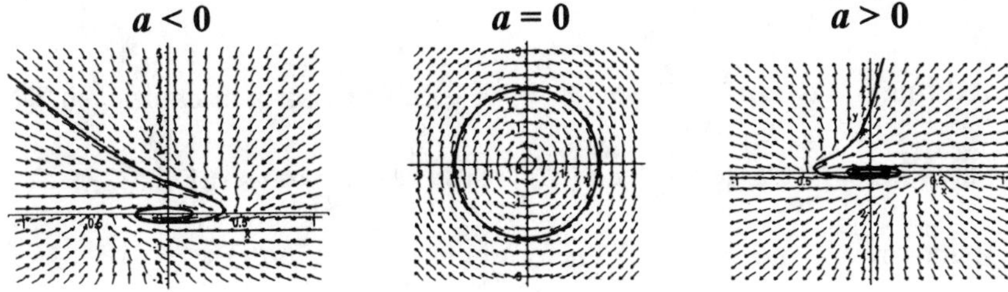

3. (b) $x' = x - y + f(x,y)$ and $y' = x + y$, where $f(x,y) = x^2$. Then $\left|\dfrac{f(x,y)}{\sqrt{x^2+y^2}}\right| = \dfrac{x^2}{\sqrt{x^2+y^2}}$

$\le \dfrac{(x^2+y^2)}{\sqrt{x^2+y^2}} = \sqrt{x^2+y^2} \to 0$ as $(x,y) \to (0,0)$.

(c) The associated linear system is $x' = x - y,\ y' = x + y$. The matrix of coefficients is $\begin{pmatrix} 1 & -1 \\ 1 & 1 \end{pmatrix}$, which has eigenvalues $1 \pm i$. Therefore, by part (A) of the Poincaré-Liapunov result and Table 5.1 on p. 241 of the text, the origin is a *spiral source*.

5. (b) $x' = -4x + y + f(x,y)$ and $y' = x - 2y + g(x,y)$, where $f(x,y) = -xy^3$ and $g(x,y) = 3x^2$.

By the Arithmetic-Geometric Mean Inequality, $|xy| \le \dfrac{x^2+y^2}{2}$, so that we have

$$\left|\frac{f(x,y)}{\sqrt{x^2+y^2}}\right| = \frac{|xy|y^2}{\sqrt{x^2+y^2}} \le \frac{(x^2+y^2)y^2\sqrt{x^2+y^2}}{2(x^2+y^2)} = \frac{y^2}{2}\sqrt{x^2+y^2} \to 0$$

as $(x,y) \to (0,0)$. Also, $\left|\dfrac{g(x,y)}{\sqrt{x^2+y^2}}\right| = \dfrac{3x^2}{\sqrt{x^2+y^2}} \le \dfrac{3(x^2+y^2)\sqrt{x^2+y^2}}{(x^2+y^2)} = 3\sqrt{x^2+y^2}$

$\to 0$ as $(x,y) \to (0,0)$.

(c) The associated linear system is $x' = -4x + y,\ y' = x - 2y$. The matrix of coefficients is $\begin{pmatrix} -4 & 1 \\ 1 & -2 \end{pmatrix}$, which has eigenvalues $-3 \pm \sqrt{2}$, both negative. Therefore, by part (A) of the Poincaré-Liapunov result and Table 5.1, the origin is a *stable node*, a *sink*.

7. (b) $x' = x - y,\ y' = 1 - e^x = 1 - \left(1 + x + \dfrac{x^2}{2!} + \dfrac{x^3}{3!} + \cdots\right) = -x - \dfrac{x^2}{2!} - \dfrac{x^3}{3!} - \cdots = -x + g(x,y)$,

where $g(x,y)=x^2\left(-\frac{1}{2!}-\frac{x}{3!}-\cdots\right)=Cx^2$. As in previous solutions, $\left|\frac{g(x,y)}{\sqrt{x^2+y^2}}\right|$ $\le C\sqrt{x^2+y^2}\to 0$ as $(x,y)\to(0,0)$.

(c) The associated linear system is $x'=x-y,\ y'=-x$. The matrix of coefficients is $\begin{pmatrix}1 & -1\\ -1 & 0\end{pmatrix}$, which has eigenvalues $(1\pm\sqrt{5})/2$. Therefore, by part (A) of the Poincaré-Liapunov result and Table 5.1, the origin is a *saddle point.*

9. (b) $x'=y+f(x,y)$ and $y'=-x+g(x,y)$, where $f(x,y)=-x^2y$ and $g(x,y)$ $=xy^2$. Then $\left|\frac{f(x,y)}{\sqrt{x^2+y^2}}\right|\le\frac{|x||xy|}{\sqrt{x^2+y^2}}\le\frac{1}{2}|x|\sqrt{x^2+y^2}\to 0$ as $(x,y)\to(0,0)$ and $\left|\frac{g(x,y)}{\sqrt{x^2+y^2}}\right|=\frac{|y||xy|}{\sqrt{x^2+y^2}}\le\frac{(x^2+y^2)|y|\sqrt{x^2+y^2}}{2(x^2+y^2)}=\frac{1}{2}|y|\sqrt{x^2+y^2}\to 0$ as $(x,y)\to(0,0)$.

(c) The associated linear system is $x'=y,\ y'=-x$. The matrix of coefficients is $\begin{pmatrix}0 & 1\\ -1 & 0\end{pmatrix}$, which has eigenvalues $\pm i$. The eigenvalues of the almost linear system are pure imaginary numbers, so the origin is either a center or a spiral point of the nonlinear system. The phase portrait of the nonlinear system indicates that the origin is a *center*:

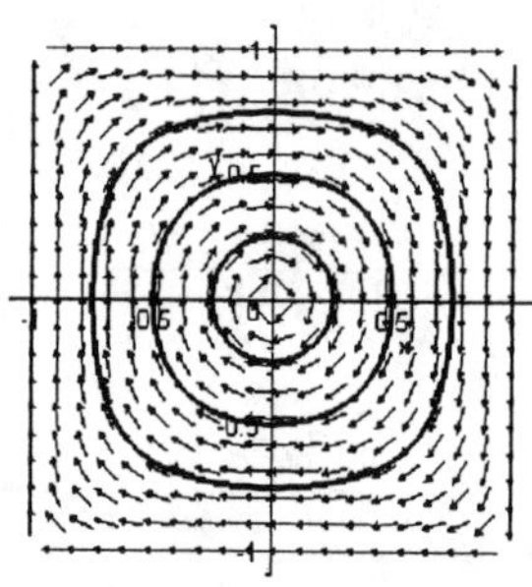

11. (b) $x'=-2x+3y+f(x,y)$ and $y'=-x+y+g(x,y)$, where $f(x,y)=xy$ and $g(x,y)=-2xy^2$.

Then $\left|\frac{f(x,y)}{\sqrt{x^2+y^2}}\right|\le\frac{|xy|}{\sqrt{x^2+y^2}}\le\frac{1}{2}\sqrt{x^2+y^2}\to 0$ as $(x,y)\to(0,0)$ and

$$\left|\frac{g(x,y)}{\sqrt{x^2+y^2}}\right|\le\frac{2|y||xy|}{\sqrt{x^2+y^2}}\le|y|\sqrt{x^2+y^2}\to 0 \text{ as } (x,y)\to(0,0).$$

(c) The associated linear system is $x'=-2x+3y,\ y'=-x+y$. The matrix of coefficients is

$\begin{pmatrix} -1 & 0 \\ 0 & -2 \end{pmatrix}$, which has eigenvalues $\left(-1 \pm \sqrt{3}\,i\right)/2$. Therefore, by part (A) of the Poincaré-Liapunov result and Table 5.1, the origin is a *spiral sink.*

13. (a) The origin is a sink. The boat eventually winds up at (0, 0). With $u = 4$ and $v = 2$, the phase portrait is

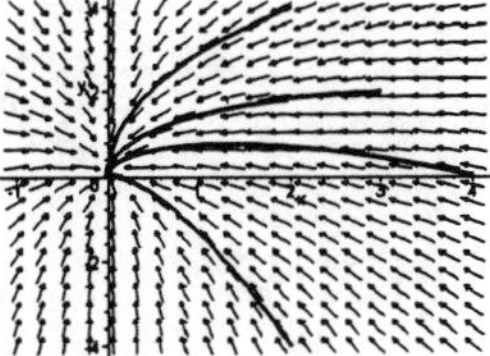

(b) The origin is now a source. The boat moves farther and farther away from (0, 0) as time passes. With $u = 2$ and $v = 4$, we get the phase portrait

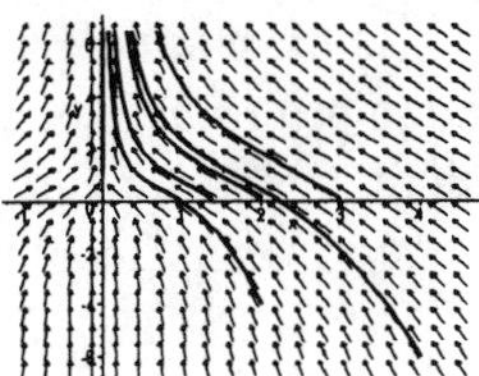

7.3 Two Important Examples of Nonlinear Equations and Systems

1. $(1)\ 3x - 2xy = 0,\ (2) - y + 4xy = 0$: $(1) \Rightarrow x(3 - 2y) = 0 \Rightarrow x = 0$ or $y = 3/2$. Now $x = 0$ in (2) $\Rightarrow y = 0$, and $y = 3/2$ in (2) $\Rightarrow x = 1/4$. Therefore, the only nontrivial equilibrium point is $\left(\frac{1}{4}, \frac{3}{2}\right)$.

3. $(1)\ 0.005x - 0.02xy = 0,\ (2) - 0.3y + 0.4xy = 0$: $(2) \Rightarrow y(-3 + 4x) = 0 \Rightarrow y = 0$ or $x = 3/4$. $y = 0$ in (1) $\Rightarrow x = 0$, and $x = 3/4$ in (1) $\Rightarrow y = 1/4$. Therefore, the only nontrivial equilibrium point is $\left(\frac{3}{4}, \frac{1}{4}\right)$.

5. A little algebra shows that $\frac{\dot{x}}{x}(x - 1) = \frac{\dot{y}}{y}(1 - y)$, or $\dot{x} - \frac{\dot{x}}{x} = \frac{\dot{y}}{y} - \dot{y}$. Integrating this equation gives $x - \ln x = \ln y - y + K$, or $x + y - \ln x - \ln y = K$. For a particular trajectory, choosing (x_0, y_0) on the trajectory determines a definite value of the constant of integration K. It is easily shown by standard calculus techniques that $f(x) = x - \ln x$ is decreasing for $0 < x < 1$ and increasing for $x > 1$. Therefore, for any given value of y there are at most two values of x giving the same value K for the expression $x + y - \ln x - \ln y$—that is, any horizontal line in the *x*-*y* phase plane intersects any particular trajectory at two or fewer points. This means that the trajectory must be a closed curve. (You should see that this proof can be modified to work in the general case when the coefficients are not all equal to one.)

7. (a) (1) $0.2x - 0.002xy = 0$, (2) $-0.1y + 0.001xy = 0$: (1) $\Rightarrow x(200 - 2y) = 0 \Rightarrow x = 0$ or $y = 100$. Now $x = 0$ in (2) $\Rightarrow y = 0$, and $y = 100$ in (2) $\Rightarrow x = 100$. Therefore, the only equilibrium points are $(0, 0)$ and $(100, 100)$.

(b) Here's the trajectory corresponding to the initial conditions $x(0) = 100$ and $y(0) = 300$:

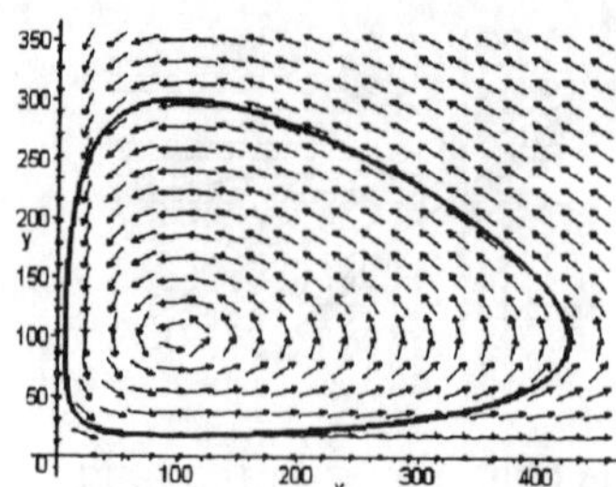

We start with a prey population of 100 and a predator population of 300, at what seems to be the highest point of the trajectory. As the number of prey declines, so does the number of predators. When the prey population begins to increase, the predator population begins to grow again. Experimentation with the right endpoint of the t-interval suggests that the period of the trajectory is about 55 (days, weeks, months,…).

(c) We can use the graph found in part (b) to estimate the maximum and minimum values of the two populations: Min $x(t) \approx 6$, Max $x(t) \approx 425$, Min $y(t) \approx 18$, Max $y(t) \approx 300$.

(d) The slope equation is $\frac{dy}{dx} = \frac{\dot{y}}{\dot{x}} = \frac{-0.1y + 0.001xy}{0.2x - 0.002xy} = \frac{y(-0.1 + 0.001x)}{x(0.2 - 0.002y)}$,
a separable equation. Separating variables, we have
$\frac{(0.2 - 0.002y)dy}{y} = \frac{(-0.1 + 0.001x)dx}{x}$, so $\int \left(\frac{0.2}{y} - 0.002\right) dy = \int \left(\frac{-0.1}{x} + 0.001\right) dx$.
This leads to $0.2 \ln y - 0.002y = -0.1 \ln x + 0.001x + C$, or
$200 \ln y - 2y = -100 \ln x + x + C$. The initial conditions yield $C = 901.27351\ldots$.
Therefore, we can write the (implicit) solution as
$x - 100 \ln x + 2y - 200 \ln y + 901.3 = 0$.

(e) Using a CAS to plot the implicit function found in part (d) for $0.1 \le x \le 450$ and $0.1 \le y \le 300$, we get

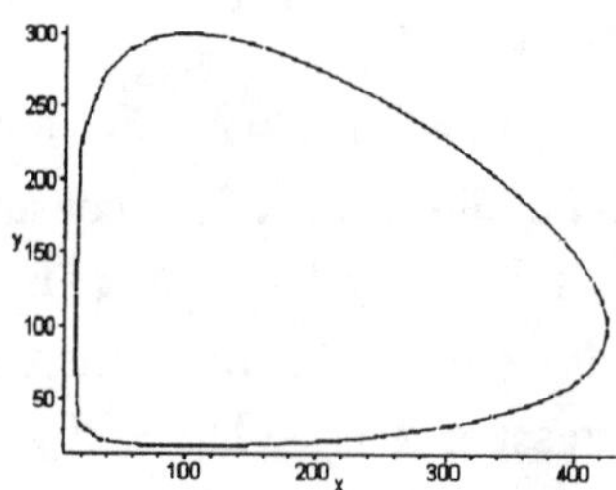

9. (a) $\dfrac{du}{dv} = \dfrac{-\left(\frac{bc}{d}\right)v}{\left(\frac{ad}{b}\right)u} = \left(-\dfrac{b^2c}{ad^2}\right)\dfrac{v}{u} \Rightarrow u\,du = -\left(\dfrac{b^2c}{ad^2}\right)v\,dv \Rightarrow u^2 = -\left(\dfrac{b^2c}{ad^2}\right)v^2 + C$

$\Rightarrow ad^2u^2 = -b^2cv^2 + K$, or $ad^2u^2 + b^2cv^2 = K$, where $K = ad^2C$ can be chosen positive.

(b) Letting $u = x - \frac{c}{d}$ and $v = y - \frac{a}{b}$, we can convert the last equation in part (a) to

$ad^2\left(x-\frac{c}{d}\right)^2 + b^2c\left(y-\frac{a}{b}\right)^2 = K$, $\dfrac{\left(x-\frac{c}{d}\right)^2}{b^2c} + \dfrac{\left(y-\frac{a}{b}\right)^2}{ad^2} = K^*$, or

$\dfrac{\left(x-\frac{c}{d}\right)^2}{\left(b\sqrt{cK^*}\right)^2} + \dfrac{\left(y-\frac{a}{b}\right)^2}{\left(d\sqrt{aK^*}\right)^2} = 1$, the standard form of the equation of an ellipse with center at $\left(\frac{c}{d}, \frac{a}{b}\right)$ and with axes parallel to the axes of the *x-y* plane. (See, for example, Section 10.2 of *Precalculus Functions and Graphs: A Graphing Approach,* Third Edition, by Larson, Hostetler, and Edwards (Boston: Houghton Mifflin Company, 2001).)

(c) $\dot{u} = (-bc/d)v$, $\dot{v} = (ad/b)u \Rightarrow \ddot{u} = (-bc/d)\dot{v} = (-bc/d)(ad/b)u = -acu$ and $\ddot{v} = (ad/b)\dot{u} = (ad/b)(-bc/d)v = -acv$.

(d) Each second-order equation found in part (c) has the form $\ddot{w} = -Rw$, or $\ddot{w} + Rw = 0$. Using the technique of Section 4.1 of the text, we find the characteristic equation $\lambda^2 + R = 0$, so $\lambda = \pm\sqrt{R}\,i$ and the general solution of the second-order differential equation is $w(t) = c_1 \cos\left(\sqrt{R}\,t\right) + c_2 \sin\left(\sqrt{R}\,t\right)$. In the case of the equations found in part (c), $R = a\,c$, so that both $u(t)$ and $v(t)$ have the form $c_1 \cos\sqrt{ac}\,t + c_2 \sin\left(\sqrt{ac}\,t\right)$. [Letting $u(t) = c_1 \cos\sqrt{ac}\,t + c_2 \sin\left(\sqrt{ac}\,t\right)$, we see that we must have $v(t) = -\dfrac{d\sqrt{ac}}{bc}\left\{-c_1 \sin\sqrt{ac}\,t + c_2 \cos\left(\sqrt{ac}\,t\right)\right)$ because $v(t) = -\left(\dfrac{d}{bc}\right)\dot{u}$.] The period of such a trigonometric expression is $\dfrac{2\pi}{\sqrt{ac}}$. [For example, see Section 4.5 of the Larson, Hostetler, Edwards book cited in part (b).]

11. Assuming that the predator population is periodic with period T (see problem 10),

$\dot{y} = -cy + dxy \Rightarrow xy = \dfrac{1}{d}(\dot{y} + cy) \Rightarrow \dfrac{1}{T}\int_0^T x(t)y(t)\,dt = \dfrac{1}{dT}\int_0^T(\dot{y} + cy)\,dt$

$= \dfrac{1}{dT}\left\{ y(T) - y(0) + c\int_0^T y(t)\,dt \right\} = \dfrac{c}{dT}\int_0^T y(t)\,dt = \dfrac{c}{d}\cdot$ (the average value of y)

= [using the result of Problem 10(b)] the average value of $x(t)$ times the average value of

$y(t) = \frac{c}{d} \cdot \frac{a}{b}.$

13. (a) $\frac{dy}{dx} = \frac{-\sin x}{y} \Rightarrow y\,dy = -\sin x\,dx \Rightarrow \int y\,dy = \int -\sin x\,dx \Rightarrow \frac{y^2}{2} = \cos x + C_1$

$\Rightarrow y^2 = 2\cos x + C \Rightarrow y = \pm\sqrt{2\cos x + C}$. If $C > 2$, $y(x)$ is periodic but never zero. The graphs of such functions are wavy trajectories such as those at the top and bottom of Figure 7.13. This corresponds to the situation described on page 337 of the text. The pendulum goes "over the top" repeatedly. A close examination of Figure 7.13 reveals that The upper waves correspond to increasing values of x and clockwise rotation of the pendulum, whereas the lower curves correspond to counterclockwise rotation.

(b) The separatrices correspond to $C = 2$. Each separatrix approaches an equilibrium point without actually reaching it. More specifically, separatrices have the odd integer multiples of π as limits on the x-axis.

15. (a) Let $x_1 = \theta$ and $x_2 = \dot{\theta}$. Then $\dot{x}_1 = y$ and $\dot{x}_2 = \ddot{\theta} = -k\dot{\theta} - \sin\theta = -k\,x_2 - \sin x_1$.

(b) The phase portrait of the damped pendulum with $k = 0.1$ is

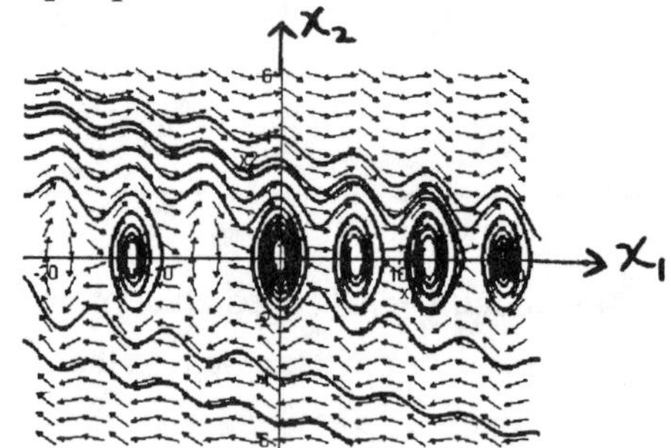

(c) The phase portrait of the damped pendulum with $k = 0.5$ is

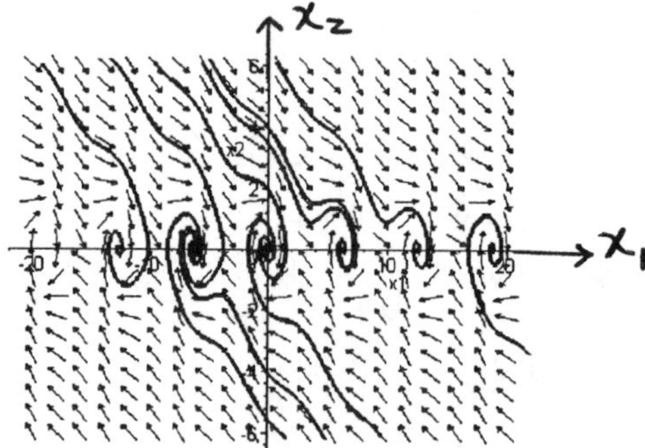

(d) The phase portrait in (b) shows that the pendulum makes a number of revolutions (dependent on the initial velocity imparted to the pendulum) and then settles into a decaying oscillation around the equilibrium point $\theta = 2k\pi$. In the phase portrait for (d), the larger coefficient of friction leads to fewer revolutions before the decaying oscillation.

7.4 Van der Pol's Equation and Limit Cycles

1. (a) When $-1 < x < 1$, we have $x^2 < 1$, so $x^2 - 1 < 0$. Because ε is a positive parameter, we see that $\varepsilon(x^2 - 1) < 0$ for $-1 < x < 1$.

   (b) When $|x| > 1$ (that is, when $x < -1$ or $x > 1$), we have $x^2 > 1$, so $x^2 - 1 > 0$. Because ε is a positive parameter, we see that $\varepsilon(x^2 - 1) > 0$ for $|x| > 1$.

3. (a) The trajectory for $\varepsilon = \frac{1}{4}$, where $x_1(0) = 1$ and $x_2(0) = 0$ is

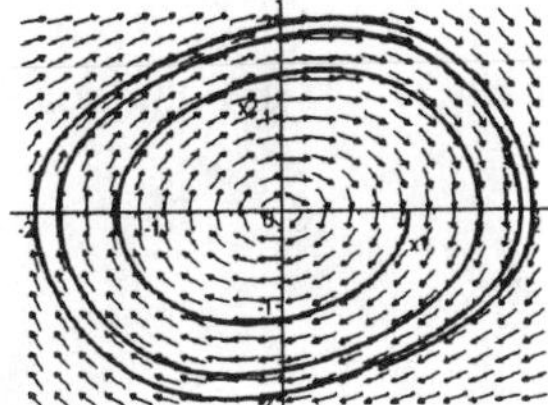

The graphs of $x_1(t)$ against t and $x_2(t)$ against t are

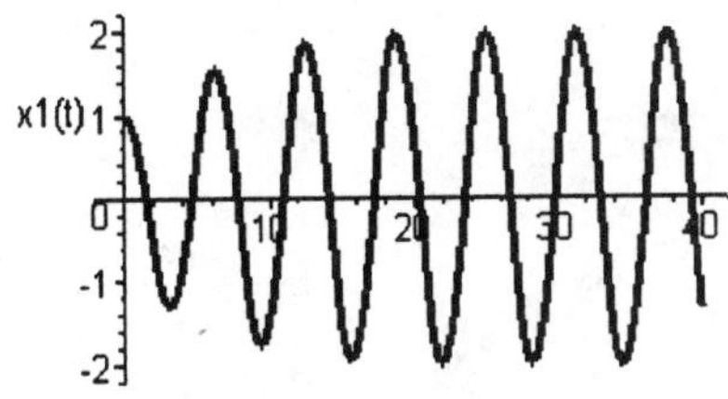

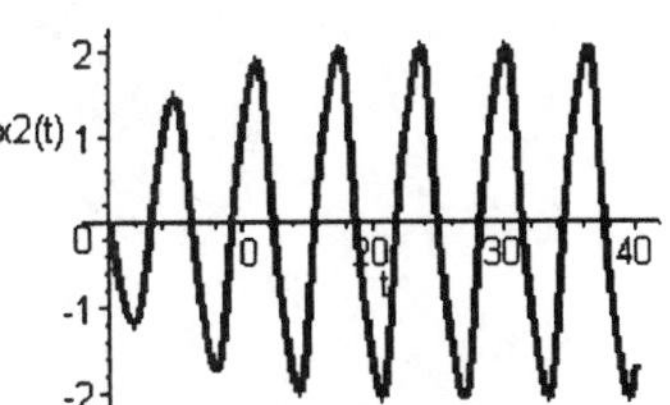

   (b) The trajectory for $\varepsilon = 4$, where $x_1(0) = 1$ and $x_2(0) = 0$ is

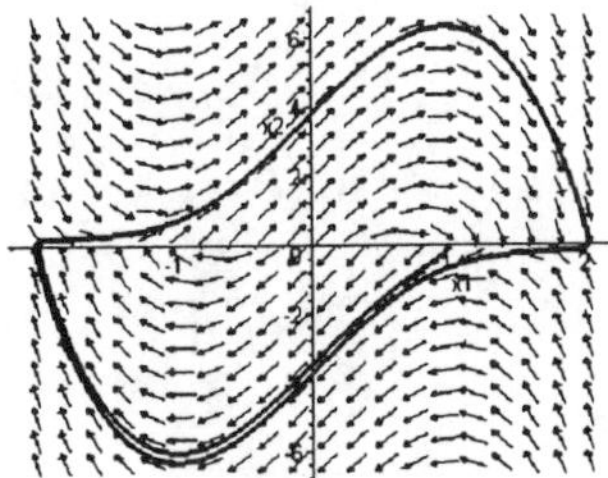

The graphs of $x_1(t)$ against t and $x_2(t)$ against t are

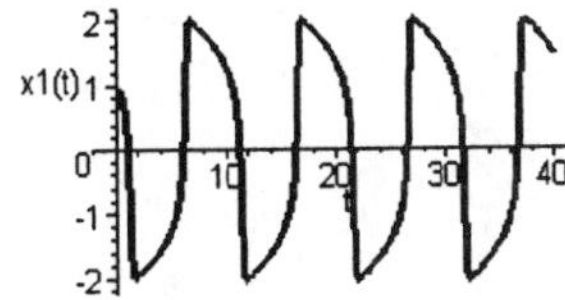

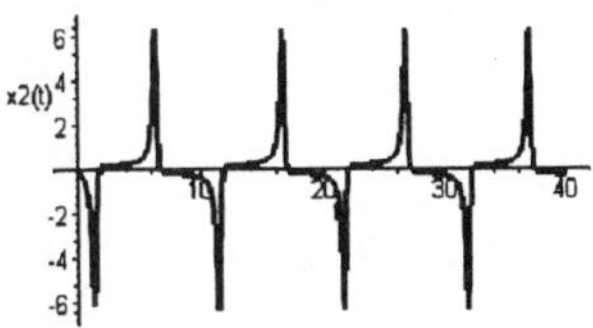

(c) Each trajectory indicates the existence of a stable limit cycle. However, the shapes of the trajectories and the limit cycles change as ε changes. Similarly, $x_1(t)$ and $x_2(t)$ are periodic but not trigonometric; and when ε changes from 1/4 to 4, $x_1(t)$ changes to a flatter shape, while $x_2(t)$ develops spikes.

5. (a) $\frac{dr}{dt} = r(r^2 - 1) \Rightarrow \left(-\frac{1}{r} + \frac{1}{2}\frac{1}{r-1} + \frac{1}{2}\frac{1}{r+1}\right) dr = dt \Rightarrow \ln\left(\frac{\sqrt{r^2-1}}{r}\right) = t + C$

$\Rightarrow \frac{\sqrt{r^2-1}}{r} = K_1 e^t \Rightarrow r(t) = \frac{\pm 1}{\sqrt{1-Ke^{2t}}}.$

(b) $\dot{\theta} = 1 \Rightarrow \theta(t) = t + C.$

(c) From parts (a) and (b), we see that $(x(t), y(t)) = (r(t)\cos\theta(t), r(t)\sin\theta(t))$

$= \left(\frac{\cos(t+C)}{\sqrt{1-Ke^{2t}}}, \frac{\sin(t+C)}{\sqrt{1-Ke^{2t}}}\right).$

(d) With a bit of skill, luck, and perseverance, you should be able to come close to Figure 7.18:

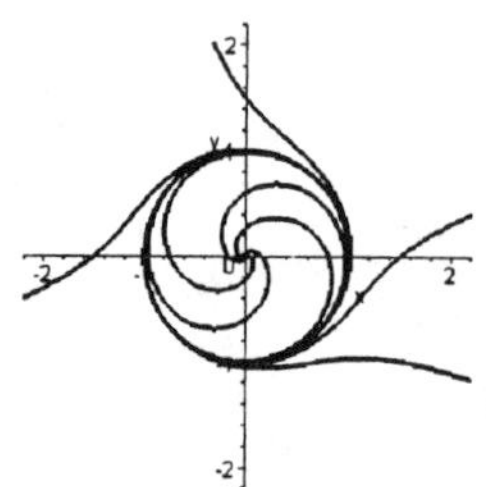

7. (a) $x(t) = r(t)\cos\theta(t),\ y(t) = r(t)\sin\theta(t),\ \dot{\theta} = 1 \Rightarrow \dot{x} = -r(\sin\theta)\dot{\theta} + \dot{r}\cos\theta$

$= -y + \dot{r}\cos\theta = -y + [r(1-r^2)]\cos\theta = -y + [r(1-r^2)]\cdot\frac{x}{r} = -y + [1-(x^2+y^2)]x$

$= x - y - x(x^2+y^2)$ and $\dot{y} = r(\cos\theta)\dot{\theta} + \dot{r}\sin\theta = x + \dot{r}\sin\theta = x + [r(1-r^2)]\sin\theta$

$= x + [r(1-r^2)]\cdot\frac{y}{r} = x + (1-r^2)y = x + [1-(x^2+y^2)]y = x + y - y(x^2+y^2).$

(b) $\dot{r} = r(1-r^2) = 0 \Rightarrow r = 0$ or $r = 1$ because $r \geq 0$. The phase portrait

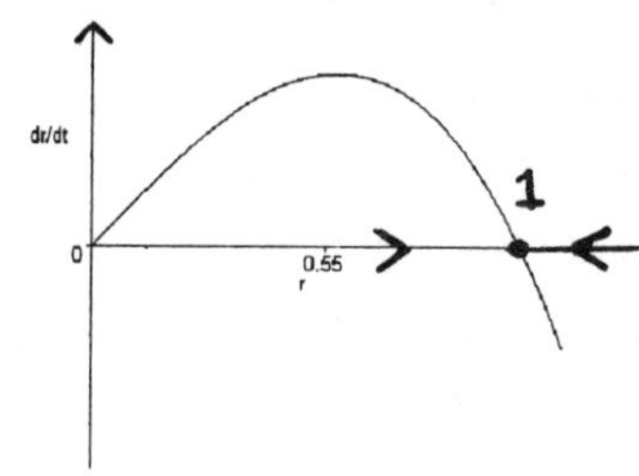

shows that the unique stable limit cycle in the unit circle, $r \equiv 1$ or $x^2 + y^2 = 1$. Trajectories starting within the unit circle approach the circumference and trajectories with initial points outside the unit circle also approach the unit circle.

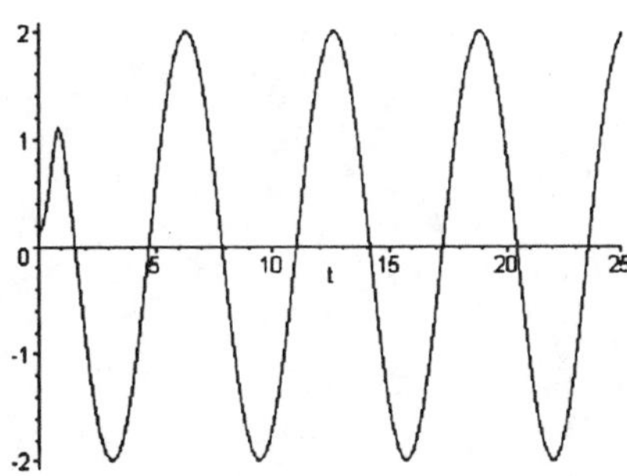

9 (a) $(1) -y-y^2=0$ and $(2)\ \frac{1}{2}x-\frac{1}{5}y+xy-\frac{6}{5}y^2=0 \Rightarrow$ [from (1)] $y(-1-y)=0$
$\Rightarrow y=0$ or $y=-1$. Now $y=0 \Rightarrow$ [from (2)] $x=0$ and $y=-1 \Rightarrow$ [from (2)] $x=-2$.
Therefore, the only equilibrium points are $(0, 0)$ and $(-2, -1)$.

(b) The phase portrait near the origin and near the point $(-2, -1)$ look like

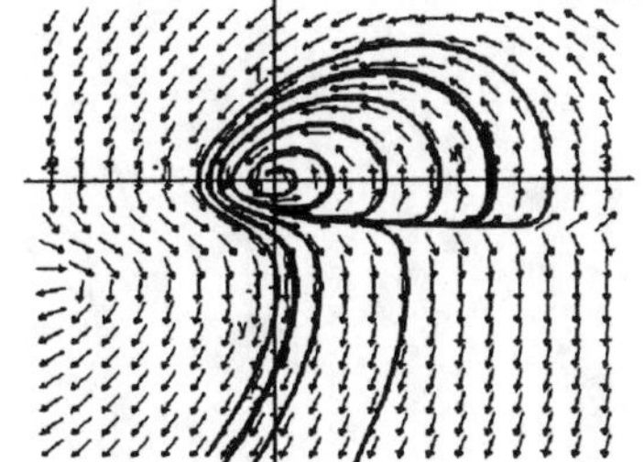

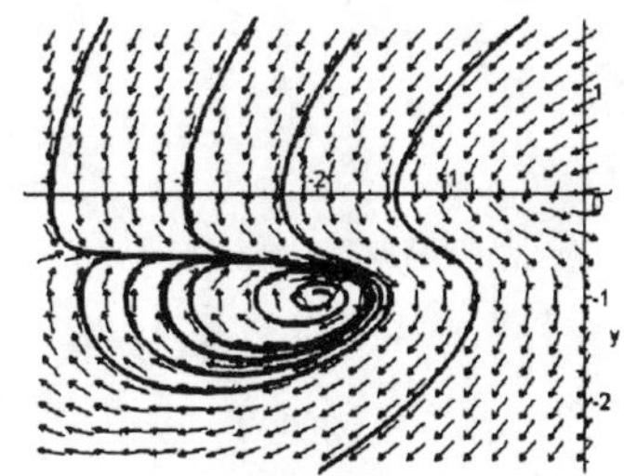

(c) There seems to be an unstable limit cycle around $(0, 0)$ and a stable limit cycle around $(-2, -1)$.

11. $\frac{\partial f}{\partial x}+\frac{\partial g}{\partial y} = (1+2y+3x^2)+(-2y+x^2) = 1+4x^2 \geq 1 > 0$ in all regions of the phase plane.

13. $\frac{\partial f}{\partial x}+\frac{\partial g}{\partial y} = (-2-\sin y) + (-3x^2y^2) < 0$ in all regions of the phase plane because
$-3 \leq -(2+\sin y) \leq -1$.

15. $\frac{\partial f}{\partial x}+\frac{\partial g}{\partial y} = (12+2xy-3x^2)+(14-2xy-3y^2) = 26-3(x^2+y^2) \geq 26-3(8) = 2 > 0$.
Therefore, by Bendixson's criterion, the system has no periodic solution in the disk
$x^2+y^2 \leq 8$.